QUALITY CONTROL ENGINEERING AND MANUFACTURING

# NATIONAL NETWORK FOR MANUFACTURING INNOVATION

# BACKGROUND AND DEVELOPMENTS

# Quality Control Engineering and Manufacturing

Additional books in this series can be found on Nova's website
under the Series tab.

Additional e-books in this series can be found on Nova's website
under the e-book tab.

QUALITY CONTROL ENGINEERING AND MANUFACTURING

# NATIONAL NETWORK FOR MANUFACTURING INNOVATION

## BACKGROUND AND DEVELOPMENTS

VIVIANNA R. BAGNOLI
EDITOR

nova publishers
*New York*

For permission to use material from this book please contact us:
Telephone 631-231-7269; Fax 631-231-8175
Web Site: http://www.novapublishers.com



Additional color graphics may be available in the e-book version of this book.

**Library of Congress Cataloging-in-Publication Data**

ISBN: 978-1-63117-681-4

*Published by Nova Science Publishers, Inc. † New York*

# CONTENTS

# PREFACE

Manufacturing plays an important role in the nation's economy, employment, and national defense. Accordingly, Congress has maintained a strong interest in the health of the U.S. manufacturing sector. The Obama Administration has undertaken a number of initiatives intended to support U.S. manufacturing, including establishment of the Advanced Manufacturing Partnership, Advanced Manufacturing National Program Office (AMNPO), Advanced Manufacturing Technology Consortia program, National Robotics Initiative, and Materials Genome Initiative. In his FY2013 budget, President Obama proposed the creation of a National Network for Manufacturing Innovation (NNMI) to help accelerate innovation by investing in industrially relevant manufacturing technologies with broad applications, and to support manufacturing technology commercialization by bridging the gap between the laboratory and the market. This book describes the National Network for Manufacturing Innovation while providing guidance on intellectual property rights for it, as well as institute performance metrics for NNMI.

Chapter 1 - Manufacturing plays an important role in the nation's economy, employment, and national defense. Accordingly, Congress has maintained a strong interest in the health of the U.S. manufacturing sector. Some analysts have expressed concerns about a decades-long decline in manufacturing employment punctuated by a steep drop from 2001 to 2010, as well as about the offshore outsourcing of production and related functions, such as research and development, by U.S. manufacturers. Others see the U.S. manufacturing sector as vibrant and healthy as evidenced by growth in output and productivity.

The Obama Administration has undertaken a number of initiatives intended to support U.S. manufacturing, including establishment of the

Advanced Manufacturing Partnership, Advanced Manufacturing National Program Office (AMNPO), Advanced Manufacturing Technology Consortia program, National Robotics Initiative, and Materials Genome Initiative.

In his FY2013 budget, President Obama proposed the creation of a National Network for Manufacturing Innovation (NNMI) to help accelerate innovation by investing in industrially relevant manufacturing technologies with broad applications, and to support manufacturing technology commercialization by bridging the gap between the laboratory and the market. Congress did not act on the President's proposal for FY2013. Nevertheless, the Administration used the Department of Defense's existing authorities and FY2012 regular appropriations to compete and award a pilot institute, the National Additive Manufacturing Innovation Institute, referring to it as "a proof-of-concept for the potential subsequent institutes."

In addition, the Administration sought nationwide input from key stakeholder groups to help guide the design of the NNMI. The AMNPO held four regional workshops and published a Request for Information (RFI) in the *Federal Register* inviting public comment on the proposed NNMI program. The input gathered from the workshops and the RFI was used by the AMNPO to prepare a National Science and Technology Committee report, *National Network for Manufacturing Innovation: A Preliminary Design,* published in January 2013.

In his 2014 budget request, President Obama again included the NNMI proposal. He also announced his intention to establish three additional manufacturing institutes in FY2013 using existing authorities and appropriations, two by the Department of Defense (DOD) and one by the Department of Energy (DOE). As in his FY2013 budget, the President's NNMI proposal calls for the establishment of up to 15 Institutes for Manufacturing Innovation (IMI) funded through a onetime infusion of $1 billion in mandatory funding to the Department of Commerce (DOC) National Institute for Standards and Technology (NIST) and to be spent over multiple years. Each IMI would be comprised of stakeholders from industry, academia, federal agencies, and state government entities. According to the proposal, each IMI is to be competitively selected, serve as a regional hub for manufacturing innovation (as well as part of the national network), and have a unique focus area (e.g., an advanced material, manufacturing process, enabling technology, or industry sector). Under the President's proposal, the NNMI would be managed collaboratively by NIST, DOD, DOE, the National Science Foundation, the National Aeronautics and Space Administration, and other agencies.

In August 2013, bills were introduced in the House (H.R. 2996) and Senate (S. 1468) to establish a Network for Manufacturing Innovation. The bills include identical provisions for establishing a Network for Manufacturing Innovation Program within the DOC's National Institute of Standards and Technology. Each bill was referred to committee(s); no further action had been taken on either bill at the time this report was published.

Chapter 2 - The Federal investment in the National Network for Manufacturing Innovation (NNMI) serves to create an effective manufacturing research infrastructure for U.S. industry and academia to solve industry-relevant problems. The NNMI will consist of linked Institutes for Manufacturing Innovation (IMIs) with common goals, but unique concentrations. In an IMI, industry, academia, and government partners leverage existing resources, collaborate, and co-invest to nurture manufacturing innovation and accelerate commercialization.

As sustainable manufacturing innovation hubs, IMIs will create, showcase, and deploy new capabilities, new products, and new processes that can impact commercial production. They will build workforce skills at all levels and enhance manufacturing capabilities in companies large and small. Institutes will draw together the best talents and capabilities from all the partners to build the proving grounds where innovations flourish and to help advance American domestic manufacturing.

Chapter 3 - Testimony of Penny Pritzker, Secretary of Commerce, U.S. Commerce Department.

Chapter 4 - This is an edited, reformatted and augmented version of a document released by the Advanced Manufacturing National Program Office, dated August 28, 2013 describing the announcement of three new institutes; the National Network for Manufacturing Innovation (NNMI), the NNMI Pilot Institute, the NNMI Institutes and evaluation and metrics.

Chapter 5 - Using a strategy of broad public engagement, in April 2012, the Advanced Manufacturing National Program Office (AMNPO) began collecting input on the National Network for Manufacturing Innovation (NNMI) program design. The collection of information from the public was initiated by a NIST Request for Information (RFI), published in the *Federal Register,* followed by a series of regional workshops sponsored by AMNPO partner agencies and focused on the issues presented in the RFI.

Reports summarizing the responses to the RFI and the comments received at each workshop were published. In January 2013, the *National Network for Manufacturing Innovation: A Preliminary Design* report was published, built upon public input received. This AMNPO document utilizes the information gathered and puts forth draft guidance associated with Intellectual Property (IP) as it relates to the proposed NNMI program, especially as it relates to the sustainability and industry impact of the individual Institutes that will comprise the NNMI.

Chapter 6 - Beginning in April 2012, a broad public engagement strategy by the Advanced Manufacturing National Program Office (AMNPO) was used to collect extensive input on the National Network for Manufacturing Innovation (NNMI or Network) program design. The collection of information from the public was initiated by a NIST Request for Information (RFI), published in the Federal Register, followed by a series of regional workshops sponsored by AMNPO partner agencies and focused on the issues presented in the RFI. Reports summarizing the findings from the RFI and each workshop were published. In January 2013, the *National Network for Manufacturing Innovation: A Preliminary Design* report was published, built upon public input received. This AMNPO document utilizes the information gathered and provides draft performance metrics that can be used to evaluate the performance of Institutes for Manufacturing Innovation (Institutes) within the NNMI program, and that would apply to the NNMI and the Institutes that are created once appropriate legislation has been enacted. This document is one of a series of documents generated to inform and seek feedback from the public on various elements that go into the creation and growth of the NNMI.

More specifically, this document describes draft metrics that can be used to help evaluate the performance and success of Institutes within the NNMI. These draft metrics were developed by an interagency team of experts, building on the initial metrics and lessons learned from the pilot institute, the National Additive Manufacturing Innovation Institute (NAMII). It is recognized that Institutes will vary considerably in their focus areas and perhaps in their structures as well; hence not all of the proposed metrics will be appropriate for all Institutes. However the six categories of draft metrics shown below are considered applicable to all Institutes:

1) Impact (1.1 – 1.4)
2) Industry Value (2.1 – 2.13)
3) Education and Workforce Development (3.1 – 3.6)
4) Portfolio (4.1 – 4.2)
5) Financial (5.1 – 5.5)
6) Network Contribution (6.1 – 6.4)

Within each broad category, several example metrics are given, and a short rationale is included for each. The examples are not intended to be comprehensive, and Institutes should be free to propose other metrics that are applicable to their particular focus areas. Among the example metrics, an effort was made to minimize the reporting burden. Some of the draft metrics are quantitative and some are qualitative. The example metrics also include items that are measurable in short, medium, or long time-frames.

In: National Network for Manufacturing … ISBN: 978-1-63117-681-4
Editor: Vivianna R. Bagnoli 

*Chapter 1*

# THE OBAMA ADMINISTRATION'S PROPOSAL TO ESTABLISH A NATIONAL NETWORK FOR MANUFACTURING INNOVATION*

*John F. Sargent, Jr.*

## SUMMARY

Manufacturing plays an important role in the nation's economy, employment, and national defense. Accordingly, Congress has maintained a strong interest in the health of the U.S. manufacturing sector. Some analysts have expressed concerns about a decades-long decline in manufacturing employment punctuated by a steep drop from 2001 to 2010, as well as about the offshore outsourcing of production and related functions, such as research and development, by U.S. manufacturers. Others see the U.S. manufacturing sector as vibrant and healthy as evidenced by growth in output and productivity.

The Obama Administration has undertaken a number of initiatives intended to support U.S. manufacturing, including establishment of the Advanced Manufacturing Partnership, Advanced Manufacturing National Program Office (AMNPO), Advanced Manufacturing Technology Consortia program, National Robotics Initiative, and Materials Genome Initiative.

* This is an edited, reformatted and augmented version of the Congressional Research Service Publication, No. R42625, dated January 29, 2014.

In his FY2013 budget, President Obama proposed the creation of a National Network for Manufacturing Innovation (NNMI) to help accelerate innovation by investing in industrially relevant manufacturing technologies with broad applications, and to support manufacturing technology commercialization by bridging the gap between the laboratory and the market. Congress did not act on the President's proposal for FY2013. Nevertheless, the Administration used the Department of Defense's existing authorities and FY2012 regular appropriations to compete and award a pilot institute, the National Additive Manufacturing Innovation Institute, referring to it as "a proof-of-concept for the potential subsequent institutes."

In addition, the Administration sought nationwide input from key stakeholder groups to help guide the design of the NNMI. The AMNPO held four regional workshops and published a Request for Information (RFI) in the *Federal Register* inviting public comment on the proposed NNMI program. The input gathered from the workshops and the RFI was used by the AMNPO to prepare a National Science and Technology Committee report, *National Network for Manufacturing Innovation: A Preliminary Design,* published in January 2013.

In his 2014 budget request, President Obama again included the NNMI proposal. He also announced his intention to establish three additional manufacturing institutes in FY2013 using existing authorities and appropriations, two by the Department of Defense (DOD) and one by the Department of Energy (DOE). As in his FY2013 budget, the President's NNMI proposal calls for the establishment of up to 15 Institutes for Manufacturing Innovation (IMI) funded through a onetime infusion of $1 billion in mandatory funding to the Department of Commerce (DOC) National Institute for Standards and Technology (NIST) and to be spent over multiple years. Each IMI would be comprised of stakeholders from industry, academia, federal agencies, and state government entities. According to the proposal, each IMI is to be competitively selected, serve as a regional hub for manufacturing innovation (as well as part of the national network), and have a unique focus area (e.g., an advanced material, manufacturing process, enabling technology, or industry sector). Under the President's proposal, the NNMI would be managed collaboratively by NIST, DOD, DOE, the National Science Foundation, the National Aeronautics and Space Administration, and other agencies.

In August 2013, bills were introduced in the House (H.R. 2996) and Senate (S. 1468) to establish a Network for Manufacturing Innovation. The bills include identical provisions for establishing a Network for Manufacturing Innovation Program within the DOC's National Institute of Standards and Technology. Each bill was referred to committee(s); no further action had been taken on either bill at the time this report was published.

# OVERVIEW

Congress maintains a strong interest in the health of the U.S. manufacturing sector due to its central roles in the U.S. economy and national defense. Manufacturing accounts for about 12% of the nation's gross domestic product (GDP) and nearly two-thirds of U.S. exports. Manufacturing enterprises directly employ nearly 12 million U.S. workers and indirectly support millions of additional jobs in other industries (e.g., banking, shipping, insurance). Manufacturers also fund about two-thirds of the nation's industrial research and development (R&D), providing a foundation for technological innovation and continued U.S. technological leadership. In addition, manufacturing workers earn higher annual wages ($47,240 in 2012) than the overall annual wages for U.S. workers ($45,790).[1] Similarly, total compensation (wages and benefits) for manufacturing workers ($79,390) exceeds total compensation for all employees ($69,710).[2] With respect to national defense, the United States depends heavily on its manufacturing base to produce the weapons, aircraft, vehicles, ships, and other equipment needed to protect the nation.

Analysts hold divergent views of the health of U.S. manufacturing. Some see the U.S. manufacturing sector as vibrant and healthy. Those holding this view tend to point to, among other things, the sector's strong growth in output and productivity, as well as the United States' world-leading share of global manufacturing output.[3] In addition, between January 2010 and November 2013, manufacturing employment added approximately 554,000 jobs, growing to more than 12.0 million.[4]

Other analysts believe that the U.S. manufacturing sector is at risk. Expressed concerns of those holding this view include:

- a "hollowing-out" of U.S. manufacturing resulting from the decision of many U.S. manufacturers to move production activities offshore and other corporate functions (e.g., research and development, accounting, information technology, tax planning, legal research);[5]
- focused efforts by other nations to grow the size, diversity, and technological prowess of their manufacturing capabilities, and to attract manufacturing operations of U.S.-headquartered multinational companies using a variety of policy tools (e.g., tax holidays, worker training incentives, market access, access to rare earth minerals); and
- a decades-long declining trend in U.S. manufacturing employment, punctuated by a steeper drop from 2001 to 2010. In January 2010,

> U.S. manufacturing employment fell to its lowest level (11.5 million) since March 1941, down more than 41% from its peak of 19.6 million in June 1979.[6]

The recent recession, relatively slow pace of recovery, and concerns about the prospects for double-dip recession[7] have contributed to increased concerns about the health of U.S. manufacturing. Some stakeholders and policy makers advocate for macro-level changes to improve the business environment, including reducing tax and regulatory burdens on manufacturers and reforming the nation's tort laws.

Others—including President Obama—support more direct and focused efforts funded by the federal government. In particular, President Obama has undertaken and proposed the creation and funding of a variety of initiatives (e.g., the Advanced Manufacturing Partnership, the National Robotics Initiative, Materials Genome Initiative) to help address concerns about U.S. manufacturing. One of the President's key proposals to help U.S. manufacturers is the establishment of a National Network for Manufacturing Innovation (NNMI).

In February 2012, the Obama Administration released *A National Strategic Plan for Advanced Manufacturing*, a report by the National Science and Technology Council (NSTC), putting forth a strategy to guide federal advanced manufacturing R&D investments. The report notes

> The acceleration of innovation for advanced manufacturing requires bridging a number of gaps in the present U.S. innovation system, particularly the gap between R&D activities and the deployment of technological innovations in domestic production of goods.[8]

The proposed NNMI seeks, in part, to bridge the innovation gap asserted in this report.

Some policy makers may oppose the NNMI for a variety of reasons, including concerns about the federal budget deficit; the appropriate role of the federal government; potential market distortions, inefficiency, and waste; and the subsidization of for-profit corporations and their shareholders at the expense of taxpayers. These issues are discussed later in this report. (See "Issues for Consideration.")

## ADMINISTRATION PROPOSAL

President Obama first proposed the establishment of a National Network for Manufacturing Innovation in his FY2013 budget, requesting $1 billion in mandatory funding[9] to support the establishment of up to 15 institutes. He formally introduced the concept on March 9, 2012, in a speech at the Rolls-Royce Crosspointe jet engine disc manufacturing facility in Virginia.

Subsequently, the Department of Defense issued a solicitation and made an award in FY2012 for "A Pilot Institute for the National Network for Manufacturing Innovation" focused on additive manufacturing. (See "The National Additive Manufacturing Innovation Institute: A Pilot NNMI Institute" later in this report for a more detailed discussion.) In addition, in the absence of congressional action on his FY2013 proposal, President Obama announced that three additional institutes would be competed and awarded in 2013—two by the Department of Defense (DOD) and one by the Department of Energy (DOE)—using existing authorities and FY2013 appropriations. (See "Affiliated Centers to be Awarded in FY2014" later in this report for a more detailed discussion.) In his FY2014 budget, President Obama once again proposed $1 billion in mandatory funding to support the establishment of the NNMI with up to 15 institutes. According to the President's proposal, the purpose of the NNMI is to bring together industry, universities and community colleges, federal agencies, and regional and state organizations

> to develop new manufacturing technologies with broad applications. Each institute will have a unique technology focus. These institutes will help support an ecosystem of manufacturing activity in local areas. The Manufacturing Innovation Institutes will support manufacturing technology commercialization by allowing new manufacturing processes and technologies to progress more smoothly from basic research to implementation in manufacturing.
>
> The NNMI Federal investment is designed to catalyze industry and non-federal co-investment in advanced manufacturing. Each institute is expected to have a plan to become self-sustaining and fully independent of NNMI Federal funds five to seven years after launch.[10]

In particular, the NNMI seeks to "advance technological innovation at a pace much faster than any one company could on its own,"[11] integrate innovation resources, improve the competitiveness of U.S. manufacturing, and encourage investment in the United States.[12] The NNMI is to be managed collaboratively by the Department of Commerce's (DOC's) National Institute of Standards and Technology (NIST), DOD, DOE, National Science Foundation (NSF), the National Aeronautics and Space Administration (NASA), and other agencies through the Advanced Manufacturing National Program Office (AMNPO),[13] a multi-agency coordination office.[14]

## Funding

As proposed in the President's FY2014 budget, NIST would receive a one-time infusion of $1 billion in mandatory funding in FY2014 to be spent over nine years (see **Table 1**). Federal funds would be used to help establish and support up to 15 Institutes for Manufacturing Innovation (IMIs, which collectively would form the NNMI) on a cost-shared basis with industrial, academic, and state and local organization partners. Each IMI is expected to become financially sustainable within seven years of its launch through income-generating activities such as member fees, intellectual property licenses, contract research, and fee-for-service activities.[15]

**Table 1. Proposed Schedule of NNMI Expenditures in millions of dollars**

| FY 2014 | FY 2015 | FY 2016 | FY 2017 | FY 2018 | FY 2019 | FY 2020 | FY 2021 | FY 2022 |
|---|---|---|---|---|---|---|---|---|
| 38 | 112 | 180 | 186 | 156 | 122 | 102 | 74 | 30 |

Source: Office of Management and Budget, Executive Office of the President, *Fiscal Year 2014 Budget of the U.S. Government*, April 2013, Table S-9, p. 203.

## Model

The NNMI is said by some to be modeled after the German Fraunhofer Institutes (see "The Fraunhofer-Gesellschaft Model" box below), which some consider to be a key facet of Germany's high-tech manufacturing success.[16] The Council on Competitiveness,[17] Information Technology and Innovation Foundation,[18] and President's Council of Advisors on Science and Technology[19] and other organizations have endorsed the NNMI concept or proposed a network of U.S.-based public-private manufacturing centers similar to the NNMI.

## The Fraunhofer-Gesellshaft Model

The Fraunhofer-Gesellschaft* is a German-based, application-oriented research organization that seeks to bridge the innovation gap that can exist between fundamental research activities and commercial applications. In particular, Fraunhofer illustrates the position of its work along the innovation path as being between Germany's Max-Planck Institutes, which conduct fundamental basic research, and industry, and its sources of funding as being balanced between public and private sources. (See graphic below.)

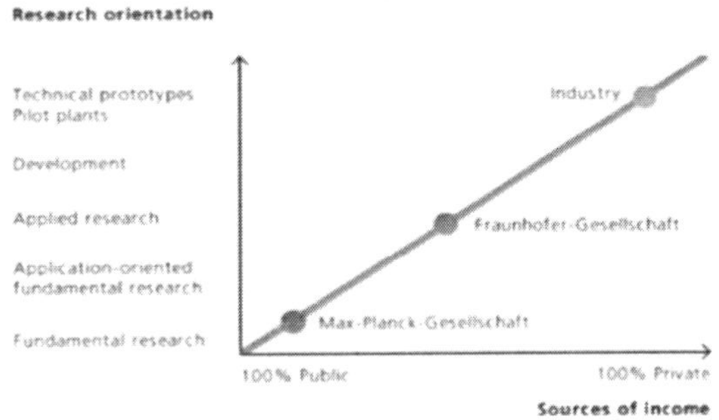


Fraunhofer has 66 institutes and research units with more than 22,000 staff members. Fraunhofer reports its annual research budget as €1.9 billion (approximately $2.5 billion). More than 70% of Fraunhofer's funding is provided through contract research with industry and publicly-financed research projects; the balance is provided by the German federal and Länder (state) governments as base funding used to support longer-term, pre-competitive research. According to Fraunhofer, the institutes bolster the competitive strength of the economy in their local regions, and throughout Germany and Europe, by promoting innovation, strengthening the technology base, improving the acceptance of new technologies, and helping to train scientists and engineers.

Fraunhofer has three subsidiaries, including Fraunhofer USA, a wholly owned non-profit organization headquartered in Plymouth, Michigan. As of 2011, Fraunhofer USA operated six centers in the United States, each working in collaboration with one or more Fraunhofer Institutes in Germany. Total 2011 revenues for Fraunhofer USA were €37.5 million (approximately $48.6 million), of which €16.1 million (approximately $20.9 million) were posted by the Fraunhofer Center for Molecular Biotechnology in Delaware, and €7.3 million (approximately $9.5 million) by the Fraunhofer Center for Sustainable Energy Systems in Massachusetts.

* "Gesellschaft" is a German word meaning association, society, or institute.

**Sources:** Fraunhofer-Gesellschaft website, http://www.fraunhofer.de; Fraunhofer *Annual Report 2011*.

## Administration's *Preliminary Design* for the NNMI

In 2012, the Advanced Manufacturing National Program Office sought nationwide input from industry, academia, state and regional governments, economic development authorities, industry associations and consortia, private citizens, and other interested parties to help guide the design of the NNMI. The AMNPO held four regional workshops and published a Request for Information (RFI) in the *Federal Register* inviting public comment on the proposed NNMI program.[20] A total of 875 people attended the workshops and the RFI drew 78 responses representing the viewpoints of more than 100 separate entities.[21]

The input gathered from the workshops and the RFI was used by the AMNPO in the preparation of a National Science and Technology Committee report, *National Network for Manufacturing Innovation: A Preliminary Design,* hereinafter the *Preliminary Design* report, published in January 2013 that articulates the Administration's perspective of the principles and characteristics for the NNMI program.[22] According to the AMNPO, the document also "builds on a review of best practices used to establish the pilot institute" (the National Additive Manufacturing Innovation Institute [NAMMI]); see "The National Additive Manufacturing Innovation Institute: A Pilot NNMI Institute" later in this report for further information.

The following sections provide an overview of selected concepts and characteristics presented in the NNMI *Preliminary Design* report.

### *Concepts*

The *Preliminary Design* report articulates a number of broad concepts for the NNMI. According to the report, the National Network for Manufacturing Innovation is to be composed of competitively selected, independently managed Institutes for Manufacturing Innovation (IMIs or institutes). Each IMI would have a specific focus area (discussed in detail below) and serve as a regional innovation hub. Collectively the institutes are to function as a network for the sharing of knowledge and best practices. The size and scale of the NNMI is intended to provide long-term regional and national economic impact.

According to the National Science and Technology Council, "A gap exists between R&D activities and the deployment of technological innovations in domestic production of goods" that is detrimental to U.S. innovation, trade, and competitiveness.[23] According to the *Preliminary Design* report, the proposed NNMI seeks to bridge this gap by strengthening support for R&D

that lies between the discovery and invention stages of innovation and the manufacturing innovation and scale up stages that precede commercialization.

The *Preliminary Design* report states that the IMIs (or institutes) are to be long-term partnerships between industry and academia (including universities and community colleges) enabled by federal, state, and local governments. The network and individual IMIs are to have a strong focus on building clusters of advanced manufacturing capabilities that join expertise from industry, academia, and government. The NNMI's emphasis is to be on linking and integrating existing public and private resources into a robust national innovation ecosystem. The IMIs are to serve as regional nodes of advanced manufacturing capabilities, where the processes to build next-generation products will be developed, demonstrated, and refined to the point where there is a clearer, lower-risk path to commercial-scale manufacturing. IMIs are to leverage existing regional or national innovation systems or catalyze the formation and sustainability of new innovation clusters. IMIs are to offer facilities comprising an "industrial commons" (the R&D, engineering, and manufacturing capabilities needed to turn inventions into competitive, manufacturable commercial products) to accelerate the formation and growth of small- and medium-sized enterprises (SMEs), and are to integrate education and workforce training functions into their operations.

Among the other concepts put forward in the *Preliminary Design* report:

- IMIs are to be based upon concepts of open innovation and partnership.
- IMIs are to facilitate the formation of effective teams of industrial and academic experts from multiple disciplines to solve difficult problems, from pre-competitive industrially relevant research to proprietary technology development for product manufacturing.
- IMIs are to provide shared-use facilities with the goal of scaling up laboratory demonstrations and maturing technologies for manufacture.
- IMIs are to create, showcase, and deploy new capabilities, new products, and new processes that can impact commercial production.
- IMIs are to engage and assist small and medium-sized enterprises in applying and adapting new manufacturing process technologies by providing technical assistance, trained personnel, and access to shared equipment and infrastructure.
- Through dual appointment of faculty and students in both research universities and IMIs, the IMIs are to develop leaders familiar with research applications, new technologies, and production systems.

- Each IMI is to have a specific physical location or locations and a clear lead organization, rather than existing as a "virtual" or distributed organization. At the same time, IMIs are to collaborate with organizations in any location that have relevant, complementary expertise.

### *Focus Areas*

With respect to focus areas, the NNMI *Preliminary Design* report states that each IMI is to have a unique and well-defined focus area, such as a manufacturing process (e.g., additive manufacturing), an enabling technology (e.g., nanotechnology), manufacturing processes for new advanced materials (e.g., carbon nanotubes), or an industry sector (e.g., medical devices). IMI focus areas are to be defined by the proposing teams, driven by the needs of industry, the opportunities created by new technologies, and the programmatic needs of the AMNPO partners.

### *Selection Process*

With respect to the process of selecting IMIs, the NNMI *Preliminary Design* report states that Institutes are to be chosen through an open, competitive application process. The solicitation and evaluation process is to be managed by the AMNPO partner agencies. Proposals are to be evaluated competitively by a review team that includes members of the AMNPO, agency partners, and other experts. The merit-based selection process may include pre-proposals, site visits, and economic and business plan analyses.

According to the *Preliminary Design* report, evaluation criteria for IMI proposals are expected to include the proposed focus area and its importance to the American economy; the plan for significant production-scale manufacturing impact in its area of specialization with respect to research, commercialization, and workforce training; the effectiveness of the governance and management structures; the proposed resources; the level of co-investment; engagement with SMEs and other community stakeholders; and the strength of the plan with respect to achieving sustainability.

### *IMI Leads and Stakeholders*

According to the *Preliminary Design* report, IMIs are to be led by an independent, U.S. not-forprofit institution with the capacity to lead an industry-wide technology, workforce development, and infrastructure agenda. Partners in the IMI are to include manufacturing enterprises of all sizes, including startups; a diverse set of institutions of higher education including both research universities and community colleges; research organizations (including federally funded research and development centers, subject to statutory or regulatory restrictions); national laboratories or government agencies (subject to funding restrictions); career and technical institutions; state, regional, and local public and private entities that support industrial clusters and associated economic development partnerships; unions; professional and industry associations; other not-for-profit organizations; and the general public. To help ensure a broad impact, IMIs are to openly encourage the addition of new partners and participants wherever relevant through well-defined mechanisms.

In addition, the report states that IMIs are envisioned as hubs linking the national and international resources that exist within the area of focus of the institute. IMIs are to leverage industry consortia, regional clusters, and other resources in science, technology, and economic development. In particular, IMIs should seek to benefit and leverage the various centers and research institutions funded through existing federal programs, such as the National Science Foundation's Engineering Research Centers (ERC) and Industry/University Cooperative Research Centers (I/UCRC) programs.

With respect to stakeholder participation, the *Preliminary Design* report states that IMIs are to demonstrate meaningful outreach to and engagement with SMEs. Each IMI is expected to engage existing intermediaries, centers, and networks that work with and address the needs of SMEs, such as the NIST Hollings Manufacturing Extension Partnership network and state and regional technology-based economic development programs. Institutes also are to be encouraged, to the extent appropriate, to seek out successful SBIR awardees to support their alignment with supply chains and explore opportunities for evolutionary demonstration of SBIR technologies within an institute's focus area.

The *Preliminary Design* report envisions IMIs assisting SMEs in a variety of ways, including, for example, information about technology trends; access to cutting edge technologies that assist with process innovations and development; shared facilities and access to specialized equipment that can accelerate product design, prototyping, and testing; and technical advice and assistance. The report also states that SMEs interested in a broad range of services and an ongoing relationship might participate in an IMI tiered membership structure that would minimize barriers to entry and encourage the membership of SMEs in the institute.

### *Foreign Participation*

The NNMI *Preliminary Design* report states that participation in an institute by a non-domestic organization is to be allowed only when in the economic interest of the United States, as demonstrated by that organization's investments in the United States in research, development, and manufacturing; significant contributions to employment in the United States; and commitment that any technology arising from or assisted by the institute be used to promote domestic manufacturing activities. Participation restrictions for non-domestic organizations may exist in some circumstances.

### *Funding*

With respect to funding, the NNMI *Preliminary Design* report notes that the President's proposal calls for a $1 billion investment from FY2014 to FY2022 to support up to 15 IMIs. Institutes are to be supported with cost-share funding from federal and non-federal sources. Each IMI is to be of sufficient size and scope to have major national and regional economic impacts and to address the multidimensional challenges associated with the institute's focus area. The level of federal funding will depend upon the magnitude of the opportunity, maturity, and capital intensity of the technology, and scope of the focus area of the institute. The report states that it expects institutes will typically receive a total of $70 million-$120 million in federal NNMI funds over five to seven years. When combined with the non-federal co-investment, the total capitalization of an IMI is expected to be in the range of $140 million to $240 million, based on a 1:1 matching of federal funding. The report states that it is likely that cooperative agreements will be the primary funding mechanism for awarding IMIs, although other types of grants and contracts may also be used.

According to the *Preliminary Design* report, each institute is to demonstrate significant co-investment support from non-federal sources during the period the institute receives federal funding. The non-federal co-investment—in cash and in-kind contributions—is to be tangible, meaningful, and in the aggregate, substantial enough to signal strong and committed industry, regional, and local partnership. In-kind and cash contributions may arise from any source, but are only counted as co-investment if they come from non-federal sources and directly support the function of the institute. The funds are to come from the IMI (e.g., revenues generated from the licensing of intellectual property); the members of the institute; state, regional, and local sources (such as economic development agencies); private donations; or other non-federal sources.

The report states that federal funding is expected to be initially larger when an institute is established and to diminish after the initial two to three years so that most of the institute's funds are provided by private and other funding sources as time progresses. Non-federal support is anticipated to be large at the time of award and primarily comprised of in-kind items such as equipment and buildings.

Continuing federal funding under the NNMI program will be contingent on co-investment by businesses and other non-federal entities in an IMI, as well as the IMI's progress toward sustainable operations. Institutes are expected to be fully independent of federal NNMI funds and sustainable within seven years of launch through income-generating activities such as membership fees, intellectual property licenses, contract research, and fee-for-service activities.

To encourage the transition to sustainability, the report states that a portion of the federal funds used for IMI projects is to be awarded competitively among the institutes. Competitive project award decisions are to be made in part based on the technical quality of proposals, as well as on prior institute performance and the strength of industrial participation.

### *Operation of the Network*

With respect to the operation of the national network of IMIs, the NNMI *Preliminary Design* states that each IMI is to communicate best practices and coordinate efforts with other institutes; coordinate approaches on issues such as intellectual property treatment, contract research, and performance metrics; be led by independent, not-for-profit institutions that coordinate industry partners both locally and nationwide, including SMEs; focus upon workforce development with its industry and academic partners at the university and community college levels to impact the engineering and technical workforce; and join in the governance and activities of an NNMI-wide Network Leadership Council.

The report states that leadership from the institutes is to formally collaborate through a Network Leadership Council made up of representatives from the IMIs, federal agencies, and other entities as appropriate. The Network Leadership Council is to oversee efforts to develop consistent and common approaches for matters such as intellectual property, contracts, research and performance metrics, and facilitating the sharing of best practices.

In addition, the *Preliminary Design* report states that:

- Each IMI's research and commercialization outcomes are to be available to other IMIs as appropriate, through technology and knowledge transfer efforts.
- To the extent possible, the institutes are to work collaboratively, sharing resources, best practices, and research and development results. In particular, IMIs are to transparently share funding and membership models, annual reports, and projections.
- Each IMI is to participate in the AMNPO-hosted Manufacturing Portal, a web-based resource to help manufacturers locate relevant research, research partners, and pertinent information within the network.

### *Activities and Operations*

With respect to the activities and operations of the NNMI, the *Preliminary Design* report states that each IMI is to focus on component validation in a relevant environment, system model or prototype demonstration in a relevant environment, system prototype demonstration in an operational environment, and actual system completion and qualification through test and demonstration.[24]

According to the report, IMI activities may include, but are not limited to:

- applied research, development, and demonstration projects that reduce the cost and risk of developing and implementing new technologies in advanced manufacturing;
- assessing the skills and certifications needed and providing educational opportunities to improve and expand the manufacturing workforce, including K-12 programs, internship opportunities, skills certification, community college engagement, university collaboration, graduate students, post-doctoral students, and retraining to meet the requirements set forth by an institute's mission in order to impact both the technical and degreed engineering workforce;
- developing innovative methodologies and practices for increasing the capabilities and capacity of supply chain expansion and integration;
- providing access to shared facility infrastructure to help reduce the cost and risk of commercializing new technologies and to address relevant manufacturing challenges on a production-level scale; and

- facilitating the creation of start-up companies to commercialize R&D results.

***Governance***

With respect to governance of the IMIs, the *Preliminary Design* report states that the interests of the three broad stakeholders of NNMI—industry, academia, and government—need to be preserved through a joint governance model. IMIs are to have substantial autonomy from its partner organizations and institutions, including an independent fiduciary board of directors predominantly composed of industry representatives and a leader in charge of day-to-day operations. IMI proposals are to outline the methods by which institute decisions will be made, including those related to operations, membership, intellectual property, capital investments, project selection, funding allocation, and sustainability.

According to the report, efficient operation of the NNMI network is to be facilitated through common policies. Common policies will facilitate interaction with SMEs, promote collaboration and movement within the network, and allow institutes to share services. While recognizing the differing needs of various manufacturing sectors, clusters, and ecosystems, the network will strive, as far as is practical, to maintain common policies with regard to intellectual property, contract research, operations, accountability, and marketing and branding.

## PRELIMINARY ACTIVITIES

The Obama Administration has undertaken efforts to lay a foundation for the NNMI in advance of possible congressional authorization and funding. In addition to the outreach effort undertaken to provide stakeholders an opportunity to help shape the design of the NNMI (see "Administration's Preliminary Design for the NNMI"), the Administration proceeded in FY2012 with the establishment of a pilot institute, the National Additive Manufacturing Innovation Institute (NAMII). Subsequently, in his February 2013 State of the Union Address, President Obama announced his intention to create three additional manufacturing innovation institutes in FY2013. The new institutes are to be established using existing FY2013 appropriations and the general authorities of the two sponsoring agencies, DOD and DOE.[25] The NAMII and the plan for establishing the new manufacturing institutes are discussed below.

## The National Additive Manufacturing Innovation Institute: A Pilot NNMI Institute

In his announcement of the NNMI in March 2012, President Obama also committed to the establishment of a pilot institute using existing resources from the Department of Defense and other federal agencies.[26] In May 2012, DOD published a broad agency announcement (BAA, a tool used for contracting) soliciting technical and cost proposals for an Additive Manufacturing Innovation Institute, describing it as "the first institute to be launched within the National Network for Manufacturing Innovation" and "a proof-of-concept for the potential subsequent institutes."[27] (The BAA described additive manufacturing as "a revolutionary suite of manufacturing technologies for building up parts, and potentially entire systems, in a layer-bylayer fashion, placing material precisely as directed by a 3D digital file.") Additive manufacturing is also sometimes referred to as "3-D printing."

On August 16, 2012, the White House announced the winning proposal, the National Additive Manufacturing Innovation Institute (NAMII), a partnership led by the National Center for Defense Manufacturing and Machining (NCDMM), based in Youngstown, OH. NAMII partners include 40 companies, 9 research universities, 5 community colleges, and 11 non-profit organizations. NAMII partners are located primarily in the Western Pennsylvania, Northeast Ohio, and Northern West Virginia region. Under the award, NAMII will receive $30 million in initial federal funding.[28] According to NDCMM, NAMII is committed to providing an additional $39 million as a cost share with funds provide mostly by industry and the states of Ohio, Pennsylvania, and West Virginia. In addition, once the institute is established it would be a "likely candidate for additional funds on a competitive basis"—up to $15 million in federal funding for specific projects.[29] These additional funds bring total potential federal funding to $45 million.

The lead agency for this institute is the Office of the Secretary of Defense (OSD), Manufacturing and Industrial Base Policy, through OSD Manufacturing Technology. The pilot institute administration is to be a cross-agency effort, primarily led by the Defense-wide Manufacturing Science and Technology Program Office, executed through the Air Force Research Laboratory. Other agencies providing funding and partnering with DOD to support NAMII include DOE, NASA, NSF, and NIST.

## Affiliated Centers to Be Awarded in FY2014

In February 2012, President Obama announced his intention to award three new manufacturing institutes using existing FY2013 funds and statutory authorities. On May 9, 2013, the White House announced competitions for two DOD manufacturing institutes—one focused on Digital Manufacturing and Design Innovation (DMDI) and the other focused on Lightweight and Modern Metals Manufacturing Innovation (LM3I)—and one DOE manufacturing institute focused on Next Generation Power Electronics Manufacturing.[30]

### *DOD Digital Manufacturing and Design Innovation Institute*

According to DOD, the DMDI Institute is expected to focus on "enterprise-wide utilization of the digital thread, enabling highly integrated manufacturing and design of complex products at reduced cost and time."[31] The core technology areas of interest include, but are not limited to, advanced manufacturing enterprise, intelligent machines, advanced analysis, and cyber-physical security. The DMDI Institute is sponsored by the Office of the Secretary of Defense and led by the Department of the Army.

The Department of the Army issued a Request for Information (RFI) for the DMDI Institute on May 9, 2013.[32] On July 5, the Army issued a Broad Agency Announcement (BAA) (BAA Announcement Number BAA-13-01DMDI) soliciting proposals for the DMDI.

### *DOD Lightweight and Modern Metals Manufacturing Innovation Institute*

The LM3I Institute is expected to focus on the integrated design and manufacturing of lightweight components and structures for commercial and defense applications, and the verification of those designs through pilot production and validation through experimental testing. The key areas of interest for the LM3I Institute are rapidly maturing and demonstrating production scale-up of existing, innovative, lightweight alloys; shortening the time necessary to design, integrate, and evaluate novel, affordable metals; developing more affordable, competitive automated manufacturing processes relevant to lightweight and modern metals; and developing the tools, skills, and knowledge base within the materials design and manufacturing workforce to use an integrated computational materials engineering infrastructure efficiently and effectively. The LM3I Institute is sponsored by the Office of Naval Research (ONR), in collaboration with the Office of the Secretary of

Defense, Deputy Assistant Secretary of Defense, Manufacturing, and Industrial Base Policy.

The ONR issued an RFI for the LM3I Institute on May 9, 2013.[33] On July 2, 2013, the ONR issued a BAA (BAA Announcement Number ONRBAA13-019) soliciting proposals for the LM3I.

***DOE Next Generation Power Electronics Manufacturing Institute***

In January 2014, the Department of Energy awarded the Next Generation Power Electronics Manufacturing Institute to a consortium of businesses and universities, led by North Carolina State University. The Next Generation Power Electronics Manufacturing Institute is being supported by the DOE Advanced Manufacturing Office[34] as a part of the department's Clean Energy Manufacturing Initiative (CEMI). The language incorporated in the DOE Funding Opportunity Announcement contained extensive language linking it to the NNMI.

The institute will focus on circuit design, packaging, module manufacturing capabilities, and wafer test metrology equipment for wide bandgap semiconductor power electronics device fabrication and manufacturing. DOE anticipates providing $14 million in initial funding, and an additional $14 million in each of the following four years for a total of $70 million.[35]

In addition to President Obama's request for $1 billion in mandatory funding for NIST in FY2014 to support the establishment of up to 15 institutes under the NNMI program, the President proposed $192.5 million in FY2014 funding for the establishment of new CEMI Institutes. The explanatory statement for Division D (Energy and Water Development and Related Agencies) of the Consolidated Appropriations Act, 2014 (P.L. 113-76) directs the Department of Energy to support the "Innovative Manufacturing Initiative to the extent possible with available funds" and "encourages research that supports development of wide bandgap semiconductor technologies."[36]

## Legislative Status

The creation of the NNMI depends on congressional authorization and funding. No action was taken by the 112$^{th}$ Congress on President Obama's FY2013 request to establish and fund the NNMI. The President renewed his call for an NNMI in his FY2014 budget, again seeking $1 billion in mandatory funding. The use of mandatory funding for the NNMI would allow funding to be provided by a law other than an appropriations bill, removing the funding decision from the regular appropriations process.[37]

In August 2013, bills were introduced in the House (H.R. 2996) and Senate (S. 1468) to establish a Network for Manufacturing Innovation. Each bill would establish a Network for Manufacturing Innovation Program within the DOC's National Institute of Standards and Technology. As introduced, each bill contains identical provisions for the establishment of the program.

The bills identify a variety of purposes to be served by the program, including:

- improving the competitiveness of United States manufacturing and increasing domestic production;
- stimulating United States leadership in advanced manufacturing research, innovation, and technology;
- facilitating the transition of innovative technologies into scalable, cost-effective, and high-performing manufacturing capabilities;
- facilitating access by manufacturing enterprises to capital-intensive infrastructure, including high-performance computing, in order to improve the speed with which such enterprises commercialize new processes and technologies;
- accelerating the development of an advanced manufacturing workforce;
- facilitating peer exchange of and the documentation of best practices in addressing advanced manufacturing challenges; and
- leveraging non-federal sources of support to promote a stable and sustainable business model without the need for long-term federal funding.

The bills would direct the Secretary of Commerce to establish a network of centers for manufacturing innovation to engage in:

- research, development, and demonstration projects, including proof-of-concept development and prototyping, to reduce the cost, time, and risk of commercializing new technologies and improvements in existing technologies, processes, products, and research and development of materials to solve pre-competitive industrial problems with economic or national security implications;
- development and implementation of education and training courses, materials, and programs;

- development of innovative methodologies and practices for supply chain integration and introduction of new technologies into supply chains;
- outreach to and engagement with small- and medium-sized manufacturing enterprises, in addition to large manufacturing enterprises; and
- other such other activities as the Secretary of Commerce, in consultation with federal departments and agencies whose missions contribute to or are affected by advanced manufacturing, considers consistent with the program's purposes.

Other provisions of the bills include:

- authorize the participation of representatives from industrial entities, research universities, community colleges, and such other entities as the Secretary considers appropriate, which may include career and technical education schools, federal laboratories, state, local, and tribal governments, businesses, educational institutions, and nonprofit organizations;
- authorize the Secretary of Commerce to award financial assistance to assist in planning, establishing, or supporting a center for manufacturing innovation;
- require the use of a competitive, merit review process in the selection of centers;
- limit financial assistance under the program to no more than seven years;
- establish a National Office of the Network for Manufacturing Innovation Program, led by a Director, to oversee and carry out the program;
- require each center receiving financial assistance from the program to report annually to the Secretary of Commerce on its expenditures and performance with respect to its goals, plans, financial support, and accomplishments, as well as to how the center has furthered the authorized purposes of the program;
- require the Secretary of Commerce to prepare an annual report to Congress on the performance of the program;
- require the Government Accountability Office to perform a triennial assessment of the program reporting on the management, coordination, and industry utility of the program; assessing the extent

to which the program has furthered its authorized purposes; and recommending legislative and administrative actions to improve the program;

- authorize the establishment of a Network for Manufacturing Innovation Fund and authorize appropriations of $600 million for the execution of the program; and
- designate the National Additive Manufacturing Innovation Institute and manufacturing centers currently under interagency review as centers for manufacturing innovation.

Following the introduction S. 1468, the bill was referred to the Senate Committee on Commerce, Science, and Transportation. Similarly, following its introduction, H.R. 2996 was referred to the House Committee on Science, Space, and Technology's Subcommittee on Research and Technology and the House Committee on Appropriations. No further action had been taken on either bill at the time this report was published.

On July 18, 2013, the Senate Committee on Appropriations reported S. 1329, the FY2014 appropriations act for the Department of Commerce, Department of Justice, and Science and Related Agencies. In the report accompanying the bill (S.Rept. 113-78), the committee asserts that there is "no significant distinction" between the proposed NNMI institutes and the Administration's Advanced Manufacturing Technology (AMTech) consortia. The report directs NIST to report within 60 days of enactment of the legislation "on how NNMI-related efforts can be merged into AmTech."[38]

## ISSUES FOR CONSIDERATION

The proposed National Network for Manufacturing Innovation raises a variety of issues for Congress, some of which were raised in a hearing on the proposal held by the House Committee on Science, Space, and Technology's Subcommittee on Technology and Innovation in May 2012. Among the questions Congress may wish to consider if it takes up legislation to establish and fund the NNMI:

- **What is the U.S. global competitive position in manufacturing?** Some assert that U.S. manufacturing is healthy and growing, pointing to indicators such as increased output and productivity; others assert that U.S. manufacturing is in decline, pointing to decreased manufacturing employment and the movement of production and related functions to other countries.

Those who see U.S. manufacturing as healthy generally assert that increased globalization and efforts to facilitate trade naturally lead companies seeking to maximize profits, open new markets, increase global market share, and better serve their customers to locate some of their production and related activities outside the United States. They also argue that, in competing against other multinational corporations, they must undertake such efforts to remain competitive. Some also argue that a variety of factors in the U.S. market (e.g., tax rates, regulations, tort law) place a heavy burden on U.S. manufacturing; in contrast, other nations may have much lower labor costs and their governments may offer a variety of incentives (e.g., tax holidays, worker training, rapid permitting) to attract and retain the manufacturing and related activities of U.S.-based companies.

Many who see U.S. manufacturing in decline assert that U.S. manufacturing capacity is being "hollowed out" as production facilities and supporting functions (such as R&D, information technology, and accounting) are sited overseas, leaving only a shell of a corporation located physically in the United States. Some experts assert that a nation's (or a state's or a region's) manufacturing strength depends on a "critical mass" of companies that are engaged in similar and supporting activities. This critical mass creates a synergy that increases the overall strength of the firms in the cluster due to a number of positive reinforcing factors (e.g., knowledge sharing, attraction of workforce talent, new start-ups, establishment of new plants, co-location of supply chains, improvements in infrastructure). As U.S.-based firms move production and related activities outside the United States, some believe that this "critical mass" may be lost, starting a downward spiral in which the synergies are lost and firms opt to move operations outside the United States where such clusters have developed.

Accordingly, many who subscribe to this view believe that manufacturing employment will decline, R&D activities will relocate to be near production facilities, and service firms that support manufacturing will be lost. In addition, some assert that the "hollowing out" will result in the loss of manufacturing capabilities needed to support the nation's military and increase the manufacturing know-how of potential adversaries.

- **Should the federal government directly or indirectly support the competitive position of the U.S. manufacturing sector? Which federal policies and programs should be prioritized? Why should the NNMI be prioritized over other approaches? Is the NNMI duplicative of other federal efforts?**

  There are many views as to what the federal government can and should do to support the competitive position of the U.S. manufacturing sector. In general, some prefer an approach that reduces costs and other burdens on manufacturers, such as reducing taxes, regulations, and frivolous lawsuits. Others prefer an expanded direct role for the federal government. This could include increasing federal funding for manufacturing R&D, providing grants and loan guarantees for domestic manufacturing, and, in some cases, subsidizing production of products for which there are deemed positive benefits for the nation that cannot be captured by the manufacturer (economists refer to such benefits as positive externalities). With a range of options that might be pursued to improve the competitive position and strength of U.S. manufacturing, some contend that the NNMI should be given high priority due to its perceived benefits (e.g., advancing research discoveries toward market-ready technologies).

  Others may believe that the NNMI is not the right approach to bolstering U.S. manufacturing. Some may assert that the role envisioned for the NNMI should be performed by the private sector; that the federal government should not favor or subsidize particular companies, industries, or technologies; that the NNMI would be ineffective or counterproductive; that the funds that would go to the NNMI should be used to support manufacturing in other ways; that the funds should be used for different federal functions altogether; or that the funds should be directed toward deficit reduction. Some may believe that the NNMI is, in part or in whole, duplicative of other federal programs, such as the Manufacturing Extension Partnership; or, as a new and separate program, represents an increasing fragmentation of federal efforts to help manufacturers. Some may question whether additional federal funding will produce more innovation and whether and how the U.S. manufacturing base will effectively absorb such innovations.

  The *Preliminary Design* report asserts that there are "notable distinctions" between the NNMI and federal manufacturing-related programs such as the NSF's Industry & University Cooperative Research (I/UCRC) and Engineering Research Centers (ERC), NIST Manufacturing Extension Partnership program, and other federal investments focused on basic and applied early stage research. Among the characteristics of the NNMI identified in the report that set it apart from other federal efforts: the scale of investment in each center, the specific focus on manufacturability and manufacturing processes and technologies; reliance on industry leadership, in partnership with other stakeholders; the significant industry co-investment which will "force [institutes] to have a pragmatic focus on industrially relevant technologies"; and the NNMI's education and workforce training approach that seeks to make community colleges an integral part of the effort. Congress may wish to explore the likely importance and effectiveness of these differentiating characteristics.

- **While the *Preliminary Design* report provides a broad outline of the functions that the network would serve and some of the roles of the federal government in the national network, other questions remain. For example: What authorities would the AMNPO have, if any, in the operation of the network? How would the federal government's role in NNMI network activity be funded? After the federal government's one-time funding is exhausted, what role would the federal government play in the network and how would this role be funded?**

  Implicit in the title of the National Network for Manufacturing Innovation is the concept of a national network. The Administration's FY2013 proposal did not discuss how the individual institutes would function as a network. The January 2014 *Preliminary Design* report[39] provided some information about some of the functions of the network and the roles of the federal government. However, it did not specifically address funding source(s) for network activities or the types of authority that the AMNPO might have in executing these roles. Congress may want to explore what form the network would take (e.g., a national network office with staff, a database of information), what functions it would perform (e.g., sharing of lessons learned, referrals of companies to centers with specific expertise), which agencies/offices would perform each of the functions (e.g., NIST, AMNPO, IMIs), and how the performance of these functions would be paid for (e.g., the NNMI mandatory appropriation, agency general appropriations, private funding from the IMIs). In addition,

Congress may opt to explore the role of the institutes in the operation of the national network.

- **Should the NNMI be funded on a mandatory or discretionary funding basis? Should a one-time advance appropriation be provided for the proposed life of the program, as proposed by President Obama, or should the NNMI be subject to annual review, oversight, and consideration in the regular annual appropriations process? Which programs would be cut or eliminated as offsets for the NNMI's proposed $1 billion mandatory appropriation?**

  The NNMI could be supported through either mandatory[40] or discretionary funding. The Obama Administration has proposed one-time mandatory funding of $1 billion to be used over the course of nine years. This approach may provide a higher degree of certainty about the availability of out-year funding for the NNMI (though Congress could opt later to rescind all or part of such funding). Such an approach would require offsets from other mandatory funding. Alternatively, providing funding through annual appropriations might allow Congress greater oversight opportunities and flexibility in modifying the program and its funding levels. If the NNMI were to be supported through discretionary funding, cuts would need to be made from other discretionary spending. The Obama Administration's budget request specified a number of cuts in mandatory spending, but did not specify which cuts would be used to offset proposed funding for the NNMI.

- **What is the appropriate role of the federal government in manufacturing-related innovation? Should the federal government's role end with basic research funding, or include funding for applied research, development efforts focused on cost reduction and technical feasibility, or demonstration projects?**

  There are many views regarding the appropriate role of the federal government in the innovation process. While there has been a general consensus on the federal government's support for basic research, congressional efforts to provide later-stage support for innovation (aside from meeting government mission requirements, such as national defense) have been met with opposition from different quarters.

- Among the arguments put forth by supporters of later-stage federal investments (e.g., applied research, development efforts, and demonstration projects):
  - In some cases, advocates assert, important benefits—for example, economic, social, national security—may be achieved that would not otherwise be achieved due to factors such as the absence of market incentives (e.g., development of drugs for diseases or conditions that only affect a small number of people) or the inability of a single company or group of companies to undertake such efforts due to high cost, high risk, and/or a long time horizon for achieving a return on investment. Benefits that cannot be captured by a company (or group of companies working together) that brings a product or service to market are referred to by economists as "positive externalities." Unable to capture these benefits, a company is not likely to consider them in its decision-making regarding whether to pursue the development of such a technology or product.

    Accordingly, economists assert that the result may be private sector underinvestment in beneficial R&D. Some analysts argue that, in such cases, public investment may be justified to induce the development of these technologies (and the realization of these benefits) by sharing costs and risks.
  - Some proponents contend that such efforts are needed to ensure U.S. leadership in technologies and industries critical to U.S. national security and economic security.
  - Some concerned about the competitiveness of U.S. industry posit that such efforts are needed to offset the industrial policies of other nations that make the business environment in the United States comparatively less attractive.

Among the arguments put forward by opponents of such efforts:

- According to some, such efforts constitute an *industrial policy*, resulting in distortions in markets and flows of capital by substituting governmental preferences for market forces.
- Some opponents argue that government is not able to make better decisions than markets, therefore federal funding for such activities is generally inefficient or wasted.

- Such efforts may, according to some, constitute a form of *corporate welfare*, providing direct or indirect public subsidies to for-profit corporations, enriching shareholders and others at the expense of taxpayers.
- Some opponents may assert that government funding and tax, regulatory, or policy decisions may be used to reward political supporters or punish opponents, referred to by some as *crony capitalism*.
- Others may assert that the ability to provide direct federal funding to companies or industries can lead to governmental corruption, fraud, and graft.
- Some oppose efforts such as the NNMI by asserting an absence of express authority in the Constitution to engage in such activities makes such efforts unconstitutional.
- Those concerned about the nation's fiscal condition argue that the current economic condition of the United States with respect to the budget deficit, national debt, and future financial liabilities does not allow for such expenditures, irrespective of merit and efficacy.

- **How can the NNMI contribute to the retention of manufacturing-related activities in the United States, both broadly as well as with respect to the R&D that the IMIs advance toward commercialization?**

  The innovation process can be extremely challenging. Even good ideas can fail due to a number of reasons (e.g., technical, cost, and risk barriers; disconnection from market needs; absence of standards; regulatory hurdles). A major thrust of U.S. science and technology policy has focused on how to move new ideas and insights from the laboratory into the marketplace.

  In the past, the strength of the U.S. economy and its position in the global economy largely meant that the success of U.S.-based companies in overcoming the obstacles to innovation and moving a process or product into the market resulted in production-related activities and jobs in the United States. Today, however, companies have increased options (and sometimes incentives) to establish production facilities outside the United States.

  The NNMI *Preliminary Design* report states that foreign organizations with significant U.S. R&D investments and employment would be allowed to participate in an IMI if they committed to U.S.-based manufacturing for any technology arising from or assisted by the institute. Congress may want to explore how effective these requirements would be in ensuring that NNMI technology and innovations are manufactured in the United States and contribute to U.S. manufacturing employment by both U.S. companies and non-domestic organizations.

- **Which agency/agencies should lead and manage the NNMI?**
  President Obama has proposed that funding for the NNMI be given to the National Institute of Standards and Technology, an agency of the U.S. Department of Commerce, through onetime mandatory funding of $1 billion to be spent over nine years. The NNMI would be managed collaboratively by NIST, DOD, DOE, NSF, and other agencies.
  As the recipient of the funds, NIST appears to be the de facto lead on the initiative. In addition, the Under Secretary for Standards and Technology and Director of NIST was the only Administration official to testify at a House hearing on the NNMI in the 112th Congress.
  NIST has played an important role in the federal government's efforts to support U.S. manufacturing since its establishment as the National Bureau of Standards in 1901. Beyond its core mission in measurement science (i.e., metrology) and standards, NIST took on its current name and was given additional authorities and programs by Congress in the late 1980s. Among these new programs were several programs focused on supporting U.S. firms, including the Advanced Technology Program (ATP, a program to accelerate the development of generic, pre-competitive, high-risk, high-payoff technologies; ATP was eliminated in 2007 and replaced by the Technology Innovation Program (TIP), which was subsequently eliminated);[41] the Malcolm Baldrige National Quality Award (a program to encourage the adoption of quality management principles by private companies and non-profit organizations); and the Manufacturing Extension Partnership (a program to enhance productivity and technological performance, and to strengthen the global competitiveness of small and medium-sized U.S.-based manufacturing firms).[42]
  The Department of Defense is also playing a key role in the NNMI. In particular, DOD awarded the National Additive Manufacturing Innovation Institute which was described in the solicitation as "the first institute to be launched within the National Network for

Manufacturing Innovation" and "a proof-of-concept for the potential subsequent institutes."[43]

Congress may wish to consider whether to designate a lead agency for the NNMI and to provide that agency with governance authorities and responsibilities, or whether to parse NNMI authorities and responsibilities among several agencies. Alternatively, Congress might opt to provide the Administration with flexibility to assign agency roles and responsibilities.

- **What requirements must be met for an TMT to be considered self-sustaining? With respect to federal policy, what would be the consequences for an TMT that does not become self-sustaining?**

As articulated by the Obama Administration, an IMI is to become self-sustaining no later than seven years from its award date. However, no definition of self-sustaining has been put forward by the Administration, nor is there any indication of consequences for failure to become self-sustaining. In addition, the Administration has stated that it expects that IMIs would be eligible to compete for funds for project specific activities under other (i.e., nonNNMI) federal programs.[44] How would such funding be considered with respect to the requirement for self-sustainability?

Federal requirements that organizations become self-sufficient have not always met with success. For example, the Omnibus Trade and Competitiveness Act of 1988 (P.L. 100-418), which established the NIST Manufacturing Extension Partnership program,[45] required centers to become self-sufficient within six years:

> In no event shall funding for a Center be provided by the Department of Commerce after the sixth year of the operation of a Center.[46]

Congress later amended the Manufacturing Extension Partnership program authorities in the Technology Administration Act of 1998 (P.L. 105-309) to allow centers to continue receiving federal funds if the center "has received a positive evaluation through an independent review," though it restricted funding after the sixth year to be no more than "one third of the capital and annual operating and maintenance costs of the Center."[47]

Congress may wish to consider whether to legislatively require that IMIs become self-sufficient, the time period within which self-sufficiency would need to be achieved, and the consequences of failing to do so.

- **What will be the role of the federal government, if any, in the NNMT after the end of the nine-year funding period?**
  As proposed, the NNMI would receive a one-time appropriation to be spent over nine years. The Administration has not articulated its vision for the NNMI past the end of this period. Congress may wish to consider what role, if any, the federal government would play in coordinating or sustaining the network after the end of this period.
- **What role, if any, should the federal government play in advancing technologies that can contribute to manufacturing competitiveness?**
  Rapid technological advances may bring revolutionary changes to the manufacturing sector in the United States and abroad. Currently, technologies such as grid computing, multi-core processors, massively parallel supercomputers, and new modeling software allow for more expansive, less expensive, and faster testing of designs. For example, automobile manufacturers are using these technologies to supplement physical crash testing of vehicles, thereby reducing costs, increasing passenger safety, and allowing for design considerations that might not otherwise have been possible.[48] According to Ford Motor Company:

> Prior to the first XJ prototype crashing into a barrier, Jaguar engineers performed more than 500 computer-simulated crash events using sophisticated crash-modeling software and this was followed up by physical tests.[49]

Further, new technologies, materials, processes, and design tools may allow for low-cost, high-customization, small lot-size production. In addition, new collaborative innovation models may open the possibility of making once-proprietary product design processes available to external creators.

These new technologies and processes—and others that are likely to emerge from global research and development efforts—may displace existing industries, companies, and workers; change where value is created in the innovation process; and affect the competitive position of nations in manufacturing. In deciding whether to authorize and fund the NNMI, Congress may also wish to consider what the appropriate role of the federal government should be, if any, in advancing the U.S. position in manufacturing broadly (i.e., what boundaries should be set, if any, to define the appropriate roles of government and the private sector).

## End Notes

[1] Occupational Employment Statistics, May 2012, Bureau of Labor Statistics, U.S. Department of Labor, http://www.bls.gov/oes.

[2] CRS analysis of data from Table 6.2D (Compensation of Employees by Industry) and Table 6.5D (Full-Time Equivalent Employees by Industry), National Income and Product Accounts, Bureau of Economic Analysis, U.S. Department of Commerce, http://www.bea.gov/iTable/iTable.cfm?ReqID=9&step=1#reqid=9&step=1&isuri=1.

[3] For more information, see CRS Report R41898, *Job Creation in the Manufacturing Revival*, by Marc Levinson.

[4] Bureau of Labor Statistics, U.S. Department of Labor, Current Employment Statistics survey database, data for manufacturing employment, all employees, seasonally-adjusted, http://data.bls.gov/cgi-bin/surveymost.

[5] For more information, see CRS Report R41712, *"Hollowing Out" in U.S. Manufacturing: Analysis and Issues for Congress*, by Marc Levinson.

[6] See footnote 3.

[7] For additional information about the recession and the subsequent pace of economic growth, see CRS Report R41444, *Double-Dip Recession: Previous Experience and Current Prospect*, by Craig K. Elwell.

[8] NSTC, Executive Office of the President, *A National Strategic Plan for Advanced Manufacturing*, February 2012, p. 1, http://www.whitehouse.gov/sites/default/files /microsites/ostp/iam_advancedmanufacturing_strategicplan_2012.pdf.

[9] Mandatory spending is controlled by laws other than appropriations acts, often through authorizing legislation. Authorizing legislation establishes or continues the operation of a federal program or agency, either indefinitely or for a specified period. In contrast, discretionary spending is provided and controlled through the annual appropriations process. For additional information on mandatory funding, see CRS Report RL33074, *Mandatory Spending Since 1962*, by Mindy R. Levit and D. Andrew Austin.

[10] *Budget of the United States Government, Fiscal Year 2014*, Appendix, p. 226, http://www.whitehouse.gov/sites/ default/files/omb/budget/fy2014/assets/com.pdf.

[11] Testimony of Patrick D. Gallagher, Under Secretary for Standards and Technology, DOC, before the U.S. Congress, House Committee on Science, Space, and Technology, Subcommittee on Technology and Innovation, *Assembling the Facts: Examining the Proposed National Network for Manufacturing Innovation*, 112th Cong., 2nd sess., May 31, 2012.

[12] NIST, "Request for Information on Proposed New Program: National Network for Manufacturing Innovation (NNMI)," *Federal Register*, Vol. 77, No. 87, pp. 26509-26511, May 4, 2012, https://federalregister.gov/a/2012-10809.

[13] The AMNPO is a multi-agency coordination office hosted by the Department of Commerce National Institute of Standards and Technology (DOC/NIST). AMNPO participating

agencies include DOC/NIST, Department of Defense (DOD), Department of Education (ED), Department of Energy (DOE), National Aeronautics and Space Administration (NASA), and the National Science Foundation (NSF).

14 *From Discovery to Scale-up: About the National Network for Manufacturing Innovation*, Advanced Manufacturing Portal, http://www.manufacturing

15 *Budget of the United States Government, Fiscal Year 2014*, Appendix, p. 226, http://www.whitehouse.gov/sites/ default/files/omb/budget/fy2014/assets/com.pdf; *National Network for Manufacturing Innovation: A Preliminary Design,* AMNPO, NSTC, Executive Office of the President, January 2013, p. ii, http://www.whitehouse.gov/sites/ default/files/microsites/ostp/nstc_nnmi_prelim_design_final.pdf.

16 U.S. Congress, House Committee on Science, Space, and Technology, Subcommittee on Technology and Innovation, *Assembling the Facts: Examining the Proposed National Network for Manufacturing Innovation*, 112th Cong., 2nd sess., May 31, 2012, http://democrats NNMI%20Hearing%20Opening%20Statement-FINAL%20WEBSITE.pdf.

17 Council on Competitiveness, *Make: An American Manufacturing Movement*, Washington, DC, December 2011, pp. 63-64, http://www.compete.org/images

18 The Information Technology and Innovation Foundation, *A Charter for Revitalizing American Manufacturing*, Washington, DC, 2011, p. 2; and "ITIF Welcomes President Obama's Proposal on Manufacturing Innovation," press release, March 9, 2012, http://www.itif.org/pressrelease/itif-welcomes-president-

19 President's Council of Advisors on Science and Technology, *Report to the President on Ensuring American Leadership in Advanced Manufacturing*, Washington, DC, June 2011, http://www.whitehouse.gov/sites/default/files/microsites/ostp/pcast-advanced-manufacturing-june2011.pdf.

20 *Federal Register*, Vol. 77, No. 87, pp. 26509-26511, May 4, 2012, https:// federalregister. gov/a/2012-10809.

21 RFI Responses & Workshop Reports, Advanced Manufacturing Portal, http://manufacturing. gov/rfi_responses.html.

22 *National Network for Manufacturing Innovation: A Preliminary Design*, AMNPO, NSTC, EOP, January 2013.

23 NSTC, Executive Office of the President, *A National Strategic Plan for Advanced Manufacturing*, February 2012, p. 1, http://www.whitehouse.gov/sites/default/files /microsites/ostp/iam_advancedmanufacturing_strategicplan_2012.pdf.

24 The activities of the institutes correspond to Technology Readiness Levels 4-7 and Manufacturing Readiness Levels 4-7, as defined in Department of Defense publications, *Technology Readiness Assessment (TRA) Deskbook* and *Manufacturing Readiness Level (MRL) Deskbook.* For additional information, see *Technology Readiness Assessment (TRA) Deskbook,* Office of the Director, Defense Research and Engineering, DOD, July 2009, http://www.skatelescope.org/public/2011-11-18_WBS-SOW_Development_Reference_Documents/ DoD_TRA_July_2009_Read_Version.pdf, and *Manufacturing Readiness Level (MRL) Deskbook, Version 2.0,* Manufacturing Technology Program, Office of the Secretary of Defense, DOD, May 2011, http://www.dodmrl.com/ MRL_Deskbook_V2.pdf.

25 In the absence of congressional authorization of the NNMI, the AMNPO is not referring to NAMII or the new centers as being IMIs within the NNMI, but rather as manufacturing institutes. The AMNPO asserts that the centers would be considered part of the NNMI, if the NNMI is authorized by Congress. Personal conversation between CRS and the AMNPO, May 2, 2013.

[26] President Barack Obama, *Remarks by the President on Manufacturing and the Economy*, The White House, Petersburg, VA, March 9, 2012, http://www.whitehouse.gov/the-press-office/2012/03/09/remarks-presidentmanufacturing-and-economy.

[27] *A Pilot Institute for the National Network for Manufacturing Innovation (NNMI)*, Solicitation Number BAA-12-17- PKM, Department of the Air Force, Air Force Materiel Command, May 8, 2012, https://www.fbo.gov/?s=opportunity& mode=form&id=2bbada5cae4ab9743 8dc3f57fed050d0.

[28] The BAA restricts the use of federal funding to direct support of the goals of the institute, for applied research, education and training, and infrastructure development, and explicitly prohibits the use of government funds to build buildings or to buy land or facilities.

[29] Testimony of Patrick D. Gallagher, Under Secretary for Standards and Technology, DOC, before the U.S. Congress, May 31, 2012.

[30] Office of the Press Secretary, The White House, "Obama Administration Launches Competition for Three New Manufacturing Innovation Institutes," press release, May 9, 2013, http://www.whitehouse.gov/the-press-office/2013/ 05/09/obama-administration-launches-competition-three-new-manufacturing-innova.

[31] The "digital thread" refers to digital information about a manufactured good that flows with it through conceptual design, detail design, manufacturing, maintenance, repair, and operation.

[32] Army Contracting Command, Department of the Army, "Request for Information (RFI) Digital Manufacturing and Design Innovation (Proposed Institute)," May 9, 2013, https:// www.fbo.gov/?s=opportunity&mode=form&id= 05872b72e4a220382fb6de25e1789db6&tab=core&_cview=0.

[33] Office of Naval Research, Department of the Navy, "Request for Information (RFI), 13-RFI-0001, Lightweight and Modern Metals Manufacturing Innovation (LM3I) Proposed Institute," May, 9, 2013, https://www.fbo.gov/?s= opportunity&mode=form&id=fa26efa 577c37ce57238aeb15bf9043a&tab=core&_cview=0.

[34] The AMO is a part of the DOE Office of Energy Efficiency and Renewable Energy (EERE).

[35] Advanced Manufacturing Office, Office of Energy Efficiency and Renewable Energy, DOE, "Funding Opportunity Announcement—Clean Energy Manufacturing Initiative," DE-FOA-0000683, CFDA Number 81.086, May 9, 2013.

[36] Explanatory statement for Division D of P.L. 113-76, http://docs.house.gov/billsthisweek /20140113/113-HR3547- JSOM-D-F.pdf.

[37] Mandatory spending is controlled by laws other than appropriations acts, often through authorizing legislation. Authorizing legislation establishes or continues the operation of a federal program or agency, either indefinitely or for a specified period. In contrast, discretionary spending is provided and controlled through the annual appropriations process. For additional information on mandatory funding, see CRS Report RL33074, *Mandatory Spending Since 1962*, by Mindy R. Levit and D. Andrew Austin.

[38] S.Rept. 113-78.

[39] NSTC, Executive Office of the President, *A National Strategic Plan for Advanced Manufacturing*, February 2012, p. 1, http://www.whitehouse.gov/sites/default/files /microsites/ostp/iam_advancedmanufacturing_strategicplan_2012.pdf.

[40] See footnote 37 for information on mandatory funding and how it differs from discretionary funding.

[41] For additional information see CRS Report 95-36, *The Advanced Technology Program*, by Wendy H. Schacht.

[42] For additional information, see CRS Report 97-104, *Manufacturing Extension Partnership Program: An Overview*, by Wendy H. Schacht.

[43] *A Pilot Institute for the National Network for Manufacturing Innovation (NNMI)*, Solicitation Number BAA-12-17- PKM, Department of the Air Force, Air Force Materiel Command, May 8, 2012, https://www.fbo.gov/?s=opportunity& mode=form&id=2bbada5cae4ab97438 dc3f57fed050d0.

[44] Testimony of Patrick D. Gallagher, Under Secretary for Standards and Technology, U.S. Department of Commerce, before the U.S. Congress. See footnote 3.

[45] P.L. 100-418 uses the term "Regional Centers for the Transfer of Manufacturing Technology" to describe what is now called the Manufacturing Extension Partnership program.

[46] P.L. 100-418.

[47] P.L. 105-309. For further discussion, see CRS Report 97-104, *Manufacturing Extension Partnership Program: An Overview*, by Wendy H. Schacht.

[48] Deborah Wince-Smith, "High Performance Computing for All," *Issues in Science and Technology*, Summer 2009.

[49] Ford Motor Company website, http://media

In: National Network for Manufacturing ... ISBN: 978-1-63117-681-4
Editor: Vivianna R. Bagnoli 

*Chapter 2*

# National Network for Manufacturing Innovation: A Preliminary Design*

## *Advanced Manufacturing National Program Office*

EXECUTIVE OFFICE OF THE PRESIDENT
**NATIONAL SCIENCE AND TECHNOLOGY COUNCIL**

January 10, 2013

Dear Colleague:

For more than a century, U.S. manufacturing has been a cornerstone of American prosperity. Today, more than ever, it is essential that the Nation continues its leadership in manufacturing—to sustain the foundations of the Nation's economic prosperity and national security and to meet new challenges in the domains of energy, transportation, health care, and education. That is one reason President Obama has set scientific and technological innovation—and support for the basic research that fuels such innovation—as central priorities of his Administration.

In furtherance of these goals, this report describes the Administration's proposed National Network for Manufacturing Innovation (NNMI)—an

* This is an edited, reformatted and augmented version of a report released by the National Science and Technology Council within the Executive Office of the President, dated January 2013.

initiative to accelerate U.S. advanced manufacturing by catalyzing the development of new technologies, educational competencies, production processes, and products via shared contributions from the public and private sectors and academia. The NNMI will consist of linked Institutes for Manufacturing Innovation (IMIs), designed to serve as regional hubs of world-leading technologies and services. IMIs will provide shared facilities to local start-ups and small manufacturers to help them scale up new technologies, accelerate technology transfer to the marketplace, and facilitate the adoption of innovative developments across supply chains. And they will act as 'teaching factories' to build workforce skills at multiple levels and to strengthen business capabilities in large and small companies.

To validate this concept, and in response to a Federal call for proposals, a public-private consortium launched a pilot IMI on Additive Manufacturing technologies in late 2012. Today that National Additive Manufacturing Innovation Institute is operating with the participation of more than 85 companies, 13 research universities, 9 community colleges, and 18 non-profit and professional associations. This report reflects lessons learned from that pilot institute as well as inputs from regional workshops and a public Request for Information.

The NNMI, which is anticipated to eventually encompass up to 15 IMIs addressing a variety of technical areas critical to U.S. global leadership, directly supports the findings of the President's Council of Advisors on Science and Technology, which has called for such a network in order to ensure a vibrant and viable American manufacturing sector for decades to come.

I look forward to working with Federal agencies, Congress, and the private and academic sectors to ensure implementation of the NNMI.

Sincerely,

John P. Holdren
Assistant to the President for Science and Technology
Director, Office of Science and Technology Policy

## EXECUTIVE SUMMARY

The Federal investment in the National Network for Manufacturing Innovation (NNMI) serves to create an effective manufacturing research infrastructure for U.S. industry and academia to solve industry-relevant problems. The NNMI will consist of linked Institutes for Manufacturing Innovation (IMIs) with common goals, but unique concentrations. In an IMI, industry, academia, and government partners leverage existing resources, collaborate, and co-invest to nurture manufacturing innovation and accelerate commercialization.

As sustainable manufacturing innovation hubs, IMIs will create, showcase, and deploy new capabilities, new products, and new processes that can impact commercial production. They will build workforce skills at all levels and enhance manufacturing capabilities in companies large and small. Institutes will draw together the best talents and capabilities from all the partners to build the proving grounds where innovations flourish and to help advance American domestic manufacturing.

Recognizing that a vibrant advanced manufacturing sector is vital to the American economy and national security, President Obama has proposed a $1 billion investment in a National Network for Manufacturing Innovation program. The NNMI program has the goal of advancing American domestic manufacturing. This program will seek to accomplish this by creating a robust national innovation ecosystem anchored by up to fifteen Institutes for Manufacturing Innovation. The Administration is committed to working with Congress to authorize and fully fund the President's request for the NNMI program. This report and the proposed program design included herein aims to support Congressional authorization and provide a guide for future program implementation.

The NNMI will fill a gap in the innovation infrastructure, allowing new manufacturing processes and technologies to progress more smoothly from basic research to implementation in manufacturing. The NNMI program has a scale and focus that is unique, and it is built upon concepts of a strong public-private partnership.

Because the challenges associated with manufacturing are multi-faceted, within the NNMI model, each IMI will have its own distinct manufacturing topic or technology focus, determined through a competitive proposal and review process managed by the AMNPO partners. This process will identify the highest value cross-cutting manufacturing challenges and opportunities, and proposals will bring together manufacturing stakeholders including government, industry, and academia. Stakeholders will consist of industry, academia (research universities and community colleges), relevant organizations (industry consortia, economic development organizations, labor organizations, national laboratories, etc.), and government bodies at all levels (Federal, State, and local). In their individual focus areas, institutes will act to anchor a region's innovation infrastructure, and will conduct research and demonstration projects. IMIs will offer facilities comprising an "industrial commons" (the R&D, engineering, and manufacturing capabilities needed to turn inventions into competitive, manufacturable commercial products) to accelerate the formation and growth of small- and medium-sized enterprises (SMEs), and will integrate education and workforce training functions into their operations. IMIs will engage with many types of corporations, with particular emphasis on engaging small and medium-sized manufacturing enterprises. They will provide shared-use facilities with the goal of scaling up laboratory demonstrations and maturing technologies for manufacture. American companies and international corporations with significant holdings in the United States are envisioned as participants in these Institutes.

Institutes will be a partnership between government, industry, and academia, supported with cost-share funding from Federal and non-Federal sources. It is expected that institutes will typically receive $70-120 million in total Federal funds, depending upon the magnitude of the opportunity, maturity, and capital intensity of the technology, and scope of the focus area, over a 5-7 year timeframe. When combined with substantial non-Federal co-investment, for example 1:1, it is envisioned the total capitalization of an institute over this period will be $140 to $240 million. At greater ratios of non-Federal co-investment, an institute could achieve its goals with a lower level of Federal funding. Institutes will be expected to be sustainable within seven years of launch through income-generating activities including member fees, intellectual property licenses, contract research, and fee-for-service activities as examples. The proposed design is of a size and scale intended to provide long-term economic impact in the region and nationally.

IMIs are to be led by independent, not-for-profit institutions that strongly leverage industry consortia, regional clusters, and other resources in science, technology, and economic development. Institutes are intended to link and leverage all available resources, including institutions funded through existing Federal programs, so that they have national and global impact.

Institutes will be established through a competitive solicitation and evaluation process managed by the interagency Advanced Manufacturing

National Program Office (AMNPO). Current participating agencies include the Departments of Commerce (DOC/NIST), Defense (DOD), Education (ED), and Energy (DOE); the National Aeronautics and Space Administration (NASA); and the National Science Foundation (NSF). Competitive solicitations will be sought and proposals will be peer-reviewed. Expected evaluation criteria include Institute focus and its importance for the American economy; the Institute plan to have significant production-scale manufacturing impact in its area of specialization from a research, commercialization, and workforce training standpoint; the effectiveness of the governance and management structures; the proposed Institute resources; the level of co-investment; engagement with SMEs and other community stakeholders; and the strength of the plan towards sustainability. Leadership from the Institutes will formally collaborate through a Network Leadership Council made up of representatives from the Institutes, Federal agencies, and other entities as appropriate. The Leadership Council will oversee efforts to develop consistent and common approaches for matters such as intellectual property, contracts, research and performance metrics, and facilitating the sharing of best practices. Each Institute will also participate in the AMNPO-hosted Manufacturing Portal, a web-based resource to help manufacturers locate relevant research, research partners, and pertinent information within the Network. Each Institute's research and commercialization outcomes will be available to other IMI's as appropriate, through technology and knowledge transfer efforts.

Beginning in April 2012, a broad public engagement strategy by the Advanced Manufacturing National Program Office was used to "crowd source" the NNMI program design initiated by a Federally sponsored Request for Information (RFI) and series of regional workshops. In parallel, a pilot institute on additive manufacturing, the National Additive Manufacturing Innovation Institute (NAMII), was announced in August 2012 to move the IMI concept forward and to refine an architecture supporting the formation of the larger Institutes and Network. The RFI period has ended, and combined with the lessons learned from NAMII and rich discussions from the regional workshops, the proposed design for the NNMI is described in this report.

## 1.0. Overview

Manufacturing plays a critical role in the American economy, underpins U.S. innovation, and is essential to national security. The health and performance of the U.S. manufacturing sector has become a topic of national interest and concern. In 2011, 11 of 19 major U.S. manufacturing industries produced less than they did in 2000 [1]. Over this same decade, more than 65,000 manufacturing establishments—nearly one in six—ceased operation [2], the United States share of global exports of advanced technology products fell from 21 percent to 15 percent [3], and the sector eliminated more than 5 million jobs.

The fundamental contributions of manufacturing to economic prosperity are evident from national and international perspectives. The U.S. manufacturing sector continues to be a mainstay of our economic productivity, generating $1.8 trillion in GDP in 2011 (12.2% of total U.S. GDP). Manufacturing firms lead the nation in exports: The $1.3 trillion of manufactured goods shipped abroad constituted 86% of all U.S. goods exported in 2011. Moreover, manufacturing has a larger multiplier effect than any other major economic activity—$1 spent in manufacturing generates $1.35 in additional economic activity [4].

Manufacturing's underpinning role also is corroborated in international studies. For example, according to the World Economic Forum, over 70 percent of the income variations of 128 nations are explained by differences in manufacturing product export [5].

U.S. manufacturing has begun to rebound from the recession of 2008 – 2010. The sector has recovered more than 500,000 jobs since January 2010 and the number of job openings for skilled factory workers has increased significantly. Still, the sector faces formidable challenges from a growing number of increasingly capable international competitors. Even as foreign nations and companies position themselves for advantage in markets for high-value-added products, the nature of manufacturing itself is undergoing transformative changes. The United States response to these challenges and changes will have significant impact on the Nation's future prosperity and its ability to meet future challenges in areas as diverse as defense, energy independence, transportation, and public health.

Numerous factors—private and public—shape the competitiveness and innovation performance of U.S. manufacturing, as documented in a stream of recent reports [6-17]. Common weaknesses identified in many of these studies might be labeled as "missed opportunities" or failures to reap the full economic and commercial value from public investments in research. There are many examples where materials and product technologies were innovated in the United States, but most of the significant commercial market share they enabled was lost to other countries. They include rechargeable lithium-ion batteries, oxide ceramics, semiconductor memory devices, manufacturing equipment such as wafer steppers, flat panel displays, robotics, solar cells, and advanced lighting.

As summed up by the National Science and Technology Council, "A gap exists between R&D activities and the deployment of technological innovations in domestic production of goods," contributing significantly, for example, to the growing trade deficit in high-value-added, advanced technology products [8].

Other nations and their manufacturing industries are investing and mobilizing resources to strengthen their manufacturing performance. Germany, Korea, and Japan each have more R&D-intensive manufacturing efforts than the United States [7]. These and other nations have established government-industryacademia partnerships to spur industrial innovation [8].

Similar large-scale public-private partnerships in the United States could accelerate the development of innovation infrastructure and strengthen the Nation's position in the global competition for new products, new markets, and new jobs. Although the United States has many elements of an innovation ecosystem, the assortment of joint centers, incubators, demonstration facilities, economic development partnerships, and other assets with innovation-related objectives tends to be fragmented and does not support an overarching U.S. manufacturing strategy.

Historically and even today, the United States has excelled at basic science and invention. But the commercial and economic rewards that can sprout and grow from these important early-stage accomplishments are realized in the stages following initial discovery—especially at the points of manufacturing scale-up and commercialization. This is especially true for complex, cost-efficient, high-value-added products that require development and mastery of equally complex manufacturing processes before successful commercialization can be achieved.

To strengthen U.S. innovation and manufacturing capacity and reverse the trend in which more research, production and design facilities, and supporting functions are sited overseas requires a robust innovation policy, as outlined in the Administration's *A National Strategic Plan for Advanced Manufacturing* [8]. A key component of this innovation policy is the creation of public-private partnerships to accelerate investment in and deployment of advanced manufacturing technologies, which the Steering Committee of the Advanced Manufacturing Partnership, composed of U.S. industry and university leaders, has recommended [7].

President Obama has proposed a $1 billion investment in a National Network for Manufacturing Innovation to advance American domestic manufacturing by addressing this gap [18]. This Federal investment will be managed collectively by several Federal agencies coordinated through the Advanced Manufacturing National Program Office (AMNPO).[†]

The NNMI program is designed to bring together industry; universities (including community colleges); and local, State and Federal governments to spur manufacturing innovation and translate it into American jobs. Up to 15 Institutes for Manufacturing Innovation will be formed.

Through long-term partnerships, the Institutes will be able to maintain a sustained focus on manufacturing technology innovation, workforce development, the transfer of promising new processes and technologies to the manufacturing sector, and the support of small- and medium-sized enterprises and aspiring start-ups for developing advanced manufacturing know-how and capabilities.

A key feature of the network and its individual IMIs is a strong focus on building clusters of advanced manufacturing capabilities that join expertise from industry, academia, and government. The NNMI will leverage existing resources with the aim of capturing the economic value of U.S. research and early-stage innovation. It will close the gap now separating American inventions, research discoveries, and ideas from the development and scale-up of domestic manufacturing capabilities necessary to make—and sell—products based on those U.S. innovations.

Emphasis will be on linking and integrating existing public and private resources into a robust national innovation ecosystem anchored by regional nodes of advanced manufacturing capabilities, where the processes to build next-generation products will be developed, demonstrated, and refined to the point where there is a clearer, lower-risk path to commercial-scale manufacturing. Institutes will facilitate and accelerate the applied and commercial development of basic research results—the outputs of work supported, for example, by the National Science Foundation or the Department of Energy's Office of Science. At IMIs, these promising "seeds" of future technologies will undergo further development and refinement into manufacturing-relevant processes and approaches for next-generation, commercially feasible products and applications that are made in the United

[†] AMNPO member agencies include the Department of Commerce and its National Institute of Standards and Technology (NIST), Department of Defense, Department of Education, Department of Energy's Advanced Manufacturing Office, NASA, and the National Science Foundation. NIST hosts the AMNPO.

States. IMIs are furthermore intended to produce benefits across agency boundaries that broadly support the competitiveness of U.S. manufacturers.

A benchmark for the NNMI institutes is the pilot institute on additive manufacturing, the National Additive Manufacturing Innovation Institute (NAMII). Launched in August 2012, NAMII is centered in Youngstown, Ohio. NAMII is composed of a broad coalition of more than 80 companies, nine research universities, six community colleges and 18 not-for-profit institutions. While it differs from the proposed NNMI and its Institutes (a single topic was selected *a priori*, and funding was based on existing programs from a number of agencies), NAMII demonstrates a number of IMI attributes. Significantly, industry and regional interests acted together to leverage the Federal investment of $30 million with a combined co-investment of $39 million. The proposed NNMI has been designed using a broad public engagement strategy, and it builds on a review of best practices used to establish the pilot institute. Public engagement was pursued through a series of workshops held nationwide as well as through comments received by the AMNPO's Request for Information, published on May 4, 2012.

The focus of each Institute will be proposed by the applicants and selected through a competitive application process. As parts of a network, each IMI will: communicate best practices and coordinate efforts with other IMI's; coordinate approaches on issues such as intellectual property treatment, contract research, and performance metrics; be led by independent, not-for-profit institutions that coordinate industry partners both locally and nationwide, including SMEs; focus upon workforce development with its industry and academic partners at the university and community college levels to impact the engineering and technical workforce; and join in the governance and activities of an NNMI-wide Network Leadership Council. The President's proposal calls for a one-time $1 billion investment over a 5-7 year period, after which each IMI should be a sustainable center of excellence.

## 2.0. THE NATIONAL NETWORK FOR MANUFACTURING INNOVATION VISION

Funded by a proposed one-time, $1 billion investment, the NNMI responds to a crucial competitiveness challenge with a vision of closing the gap between research and development (R&D) activities and the deployment of technological innovations in the domestic production of goods.

The proposal implements recommendations made by the President's Council of Advisors on Science and Technology (PCAST) and a wide range of other experts and organizations. It recognizes that investing in early stage research isn't enough to ensure that a new technology progresses smoothly from invention to product development to manufacturing at scale in the United States.

The IMIs will bring together industry, universities and community colleges, Federal agencies, States, and localities to accelerate manufacturing innovation and scale up by investing in industrially-relevant, cross-cutting product and process technologies. These stakeholders will co-invest with the Federal Government in each IMI, forming a strong partnership between industrial and other stakeholders to catalyze a renaissance in advanced manufacturing activity.

IMI activities may include, but are not limited to: applied research and demonstration projects that reduce the cost and risk of commercializing and manufacturing new technologies or that solve generic manufacturing problems; education and training at all levels; development of innovative methodologies and practices for enhancing the capabilities of and integrating supply chains; and engagement with small and medium-sized manufacturing enterprises (SMEs).

The President's proposed NNMI and the regional collaborations it catalyzes are designed to address barriers to rapid and efficient development and commercialization of new advanced manufacturing-process innovations, thus strengthening existing or building new innovation ecosystems in advanced manufacturing. The network and its individual IMIs will help companies to collaborate and access the capabilities of research universities and other science and technology organizations to support efforts aimed at developing and scaling up manufacturing and assembly processes. At the same time, the IMIs will help to meet the challenge of building at all levels the pool of high-skilled talent that advanced manufacturing innovation and production requires. The number of proposed IMIs, envisioned to be up to 15 institutes, will be scaled in order to provide sufficient resources for achieving regional and national impacts in advanced-manufacturing performance.

The NNMI has two fundamental components: the individual institutes and the coordinating network, as described below.

# 3.0. CHARACTERISTICS OF THE INSTITUTES FOR MANUFACTURING INNOVATION

The Federal investment in the National Network for Manufacturing Innovation (NNMI) serves to create an effective manufacturing research infrastructure for U.S. industry and academia to solve industry-relevant problems. The NNMI will consist of linked Institutes for Manufacturing Innovation (IMIs) with common goals, but unique concentrations. In an IMI, industry, academia, and government partners leverage existing resources, collaborate, and co-invest to nurture manufacturing innovation and accelerate commercialization.

As sustainable manufacturing innovation hubs, IMIs will create, showcase, and deploy new capabili ties, new products, and new processes that can impact commercial production. They will build workforce skills at all levels and enhance manufacturing capabilities in companies large and small. Institutes will draw together the best talents and capabilities from all the partners to build the proving grounds where innovations flourish and to help advance American domestic manufacturing.

Broad public input was received regarding the desirable characteristics of IMIs and the Network as a whole through the four public "Designing for Impact" workshops held in Troy, New York; Cleveland, Ohio; Irvine, California; and Boulder, Colorado. In addition, substantial input was received through the RFI responses collected by the AMNPO. Detailed information about the participants in the workshops and respondents to the RFI may be found in Appendix A. In total, comments from nearly 900 stakeholders have contributed to the design of the Institutes and Network. AMNPO has reviewed and distilled all of this input for incorporation into this document.

IMIs will be long-term partnerships between industry and academia (including universities and community colleges) enabled by Federal, State, and local governments. In order to advance American domestic manufacturing, they will have a sustained focus on manufacturing technology innovation with a strong brand identity and reputation. They will identify critical manufacturing processes and technologies with potential transformational impact, and through their member companies, they will have the capacity to translate these technologies into market-relevant private-sector manufacturing production. IMIs will facilitate the formation of effective teams of industrial and academic experts from multiple disciplines to solve difficult problems, from pre-competitive industrially relevant research to proprietary technology development for product manufacturing. Through dual appointment of faculty and students in both research universities and Institutes, they will develop leaders familiar with research applications, new technologies, and production systems. They will engage and assist SMEs in applying and adapting new process technologies by providing technical assistance, highly trained personnel, and access to shared equipment and infrastructure. IMIs will provide education and training opportunities to build and enhance the skills of the manufacturing workforce.

## 3.1. Overall IMI Focus

The focus of each Institute will be on integrating capabilities through collaborations at facilities designed and equipped to further efforts addressing cross-cutting manufacturing challenges, yielding solutions that have the potential to retain or expand industrial production in the United States.

### *3.1.1. Defining Focus Areas*

Each Institute will have a unique and well-defined focus area, such as a manufacturing process, an enabling technology, manufacturing processes for new advanced materials, or an industry sector (see Section 3.1.2). There was broad consensus among the workshop participants and RFI respondents that the emergent focus areas should be defined by the proposing teams. The proposing teams will be driven by the needs of industry, the opportunities created by new technologies, and the programmatic needs of the AMNPO partners. Many respondents cautioned against the tendency to pre-select trendy topics that don't serve this essential balance. This concept, strongly supported from public input, is to allow open solicitations with clear selection criteria rather than government-selected topics. Good ideas come from unexpected places. The government view is included as the agencies will make the selections.

The focus area for each Institute will be defined by the proposing team, and the AMNPO partners will evaluate the efficacy of the proposal in meeting a national need as part of an overall portfolio. Proposing teams will need to demonstrate that their focus area is appropriate for an Institute, including the potential to deliver regional and national improvements in advanced-manufacturing capabilities, and to meet national needs. IMIs will provide a comprehensive and detailed plan to accomplish institute objectives and impacts.

Each Institute should leverage existing regional or national innovation ecosystems or catalyze the formation and sustainability of new innovation clusters. Each Institute will have a specific physical location or locations and a clear lead organization, rather than existing as a "virtual" or distributed organization.

At the same time, the Institute will collaborate with organizations in any location that have relevant, complementary expertise. It is expected that Institutes will take advantage of technology clusters, and will draw talent and resources from across the nation and around the world to their site.

### *3.1.2. Illustrative Focus Areas*

For the purpose of illustration only, the following are examples for each of the categories of Institute focus areas:

1) **Manufacturing Process:** Refining standards, materials, and equipment for additive manufacturing to enable low-cost, low-volume production using digital designs that can be transmitted from designers located anywhere. This focus area was selected for the pilot National Additive Manufacturing Innovation Institute, which uses existing resources and authorities. Alternate examples include advanced joining or polymer processing methods.
2) **Advanced Materials:** Developing lower-cost production methods for lightweight materials, such as low-cost carbon fiber composites, for applications that will improve fuel efficiency, performance, and corrosion resistance of the next generation of automobiles, aircraft, ships, and trains. Alternate examples include novel materials development for improved manufacturability at scale in solar power or next-generation integrated circuits.
3) **Enabling Technology:** Creating a smart-manufacturing infrastructure and approaches that integrate low-cost sensors into manufacturing processes, enabling operators to make real-time use of "big data" flows from fully instrumented plants in order to improve productivity, optimize supply chains, reduce costs, and reduce energy, water, and materials consumption.
4) **Industry Sector:** Improving medical device or biomaterials manufacturing processes to enhance safety, quality, and consistency of pharmaceuticals, chemicals, or other products by enabling rapid on-line sensing and analytical capabilities, and creating new tools for bioprocess optimization and control to enable cost-effective

production methods. Other examples are development of next-generation automotive or aerospace manufacturing process technologies.

## 3.2. IMI Activities

According to the National Science Foundation [19], approximately 17.4 percent of total public and private R&D spending in the United States supported basic research, 22.3 percent supported applied research, and 60.3 percent supported development in 2008. Within these categories the Federal Government continues to be the nation's (and the world's) largest funder of basic, knowledge-driven R&D. Approximately 21 percent of Federal R&D supports basic research, while 22 percent supports applied research [20]. While U.S. industry accounts for almost two-thirds of all U.S. R&D funding, about 90 percent of the industry investment supports applied research and technology demonstration efforts with low levels of technical and commercial risks and short payback periods [7]. The NNMI proposal is aimed at strengthening support for R&D that lies between the 'discover/invent' beginnings of innovation and the 'manufacturing innovation/scale up'stages that precede commercialization.

The Institutes, therefore, will focus on Technology Readiness Levels 4-7, defined in Table 1, which include component validation in a relevant environment, system model or prototype demonstration in a relevant environment, system prototype demonstration in an operational environment, and actual system completion and qualification through test and demonstration. The NNMI will launch a government-industry-academia partnership that galvanizes the resources of all stakeholders to achieve the critical mass of efforts needed to effectively address these stages of manufacturing innovation.

IMI activities will include, but are not limited to:

1) Applied research, development, and demonstration projects that reduce the cost and risk of developing and implementing new technologies in advanced manufacturing. An Institute's research activities will be driven by the need to mature the demonstration of component technologies in a laboratory environment toward the demonstration of a system in a representative production environment.

2) Education and training at all levels. An Institute will assess skills and certifications needed and provide educational opportunities to improve and expand the manufacturing workforce, including K-12 programs, internship opportunities, skills certification, community college engagement, university collaboration, graduate students, post-doctoral students, and retraining to meet the requirements set forth by an Institute's mission in order to impact both the technical and degreed engineering workforce.
3) Development of innovative methodologies and practices for increasing the capabilities and capacity of supply chain expansion and integration. Deployment of these will be facilitated through the Institute and its strategic partnerships and efforts to impact the larger industrial complex.
4) Engagement with SMEs. Small and medium-sized enterprises are a vital part of the manufacturing sector; SMEs in highly effective supply chains for technology-intensive manufacturing sectors have the potential to accelerate efforts to enhance U.S. competitiveness in advanced manufacturing overall. These SMEs tend to be early adopters of transformational technology, and they are well-positioned to innovate and produce jobs. This makes their involvement in Institutes essential for maximizing industry and economic impacts. However, SMEs usually invest less in R&D than larger or more established firms. Strategies to encourage the participation of SMEs in Institutes include engaging outreach partners and intermediaries that work closely with SMEs, providing valued information and services tailored to address SME needs, providing a tiered membership fee structure, allowance of all in-kind contributions for new member SMEs, the use of contract-based activities, staged licensing of IP, and similar arrangements.
5) Shared facility infrastructure. Institutes will provide shared facilities to local industry, especially SMEs and startups, with the goal of scaling up laboratory demonstrations and making technologies ready for manufacture. Each IMI will integrate capabilities and facilities required to reduce the cost and risk of commercializing new technologies and to address relevant manufacturing challenges on a production-level scale.

**Table 1. Technology Readiness Levels and Manufacturing Readiness Levels, after [21]**

| | | | | |
|---|---|---|---|---|
| | TRL 1: | Basic principles observed and reported | MRL 1: | Manufacturing feasibility assessed |
| | TRL 2: | Technology concept and/or application formulated | MRL 2: | Manufacturing concepts defined |
| | TRL 3: | Analytical and experimental critical function and/or characteristic proof of concept | MRL 3: | Manufacturing concepts developed |
| NNMI Target | TRL 4: | Component and/or breadboard validation in a laboratory environment | MRL 4: | Capability to produce the technology in a laboratory environment |
| | TRL 5: | Component or breadboard validation in a relevant environment | MRL 5: | Capability to produce prototype components in a production relevant environment |
| | TRL 6: | System/subsystem model or prototype demonstration in a relevant environment | MRL 6: | Capability to produce prototype system or subsystem in a production relevant environment |
| | TRL 7: | System prototype demonstration in an operational environment | MRL 7: | Capability to produce systems, subsystems or components in a production relevant environment |
| | TRL 8: | Actual system completed and qualified through test and demonstrated | MRL 8: | Pilot line capability demonstrated; Ready to begin Low Rate Initial Production |
| | TRL 9: | Actual system proven through successful mission operations | MRL 9: | Low rate production demonstrated; Capability in place to begin Full Rate Production |

## 3.3. IMI Funding, Revenue, and Sustainability

Each Institute should be of sufficient size and scope to have major national and regional economic impacts and to address the multidimensional challenges associated with the Institute's focus area. The amount of Federal funding should be appropriate to the Institute proposed. Federal funding to launch a typical Institute is expected to range from $70-$120 million over a 5-7 year timeframe. With greater than 1:1 non-Federal co-investment, an institute could achieve its goals with a lower level of Federal funding. In addition, each Institute should build into its operations model the ability to develop and leverage diverse revenue streams throughout its life. Federal funding is expected to be initially larger when an Institute is established and to diminish after the initial 2-3 years so that most of the Institute's funds are provided by private and other funding sources as time progresses.

Each Institute should have a plan to be self-sustaining, based on diverse funding sources beginning at Institute formation, and be fully independent of NNMI Federal funds 5-7 years after launch. Institutes will have the flexibility to pursue sustainable Institute revenue from a variety of sources including: member fees, fee-for-service activities, contract research or pre-production scale-up, non-NNMI grants and awards from Federal and other sources, intellectual property royalties, endowments, etc. The Board of Directors for each Institute will develop, within constraints specified by the Institute award, policies and procedures for its operation and for its revenue-generating mechanisms. Institute facilities and work products should be made available on appropriate and reasonable terms to a broad base of industrial users. The evolution of Institute facilities and work activities over time should also be informed by a broad base of industrial partners and other stakeholders.

To encourage the transition to sustainability, a portion of the Federal funds used for Institute projects will be awarded competitively among the Institutes. Competitive Project Award decisions will be made in part based on technical quality, but additionally based on prior Institute performance and the strength of industrial participation. This is described further in Section 3.7.

During the period of NNMI Federal funding, each Institute should demonstrate significant co-investment support from non-Federal sources. The non-Federal co-investment must be tangible, meaningful, and in the aggregate, substantial enough to signal strong and committed industry, regional, and local partnership. The co-investment requirement is expected to be met with both cash and in-kind contributions. In-kind and cash contributions may arise from any source, but are only counted as co-investment if they come from non-

Federal sources and directly support the function of the Institute. The funds may come from the Institute (for example, using revenue from the licensing of intellectual property); the members of the Institute; State, regional, and local sources (such as economic development agencies); private donations; or other non-Federal sources. The value of in-kind contributions should be determined in accordance with OMB Circular A-110, Subpart C, Section 23.

## 3.4. IMI Partners and Members

The acceleration of innovation for advanced manufacturing in order to advance American domestic manufacturing requires bridging a number of gaps in the present U.S. innovation system. Each Institute will be based upon concepts of open innovation and partnerships.

Institutes should be led by an independent, U.S. not-for-profit institution with the capacity to lead an industry-wide technology, workforce development, and infrastructure agenda. Although other lead organizations (such as universities) are not prohibited, many workshop participants and RFI respondents expressed concern about the treatment of intellectual property, confidentiality, tax-exempt status, and timely responses if universities were allowed to lead Institutes. Partners in the Institute should include the full range of national, State, and local stakeholders. This includes manufacturing enterprises of all sizes including startups. In addition, a diverse set of institutions of higher education including both research universities and community colleges should be included. Other partners may include research organizations (including Federally Funded Research and Development Centers, subject to statutory or regulatory restrictions); national laboratories or government agencies (subject to funding restrictions); career and technical institutions; State, regional, and local public and private entities that support industrial clusters and associated economic development partnerships; unions; professional and industry associations; other not-for-profit organizations; and the general public. To help ensure a broad impact, Institutes should openly encourage the addition of new partners and participants wherever relevant through well-defined mechanisms.

To effectively leverage existing national capabilities and centers of excellence, Institutes are envisioned to be hubs that link the national and international resources that exist within the area of focus of the Institute. To this end, Institutes should seek to benefit and leverage the various centers and research institutions funded through existing Federal programs, such as the

NSF Engineering Research Centers (ERC) and Industry/University Cooperative Research Centers (I/UCRC) programs. These programs focus teams of faculty and students on research that provides opportunities to advance technology for their member firms through formalized partnerships with industry. The ERCs have education programs that create pathways to engineering for pre-college students. These programs also support pre-college teachers to learn engineering concepts at the centers and develop course materials to bring engineering to pre-college classrooms.

Participation in an Institute by a non-domestic organization will be allowed only when in the economic interest of the United States. This would be demonstrated by that organization's investments in the United States in research, development, and manufacturing; significant contributions to employment in the United States; and commitment that any technology arising from or assisted by the Institute be used to promote domestic manufacturing activities. Participation restrictions for non-domestic organizations may exist in some circumstances.

## 3.5. Engagement of IMIs with Small and Medium-sized Enterprises

A major goal of the NNMI is to support the creation, growth, and expansion of domestic SMEs, and to that end, each Institute will demonstrate meaningful outreach to and engagement with SMEs. Each Institute is expected to engage existing intermediaries, centers, and networks that work with and address the needs of SMEs, to the benefit and success of the Institute's advanced manufacturing agenda. Specific examples of organizations that, through integration with the Institute, would strengthen and expand involvement of SMEs include the NIST Hollings Manufacturing Extension Partnership (MEP) network and State and regional technology-based economic development programs. Institutes will also be encouraged, to the extent appropriate, to seek out successful SBIR awardees to support their alignment with supply chains and explore opportunities for evolutionary demonstration of SBIR technologies within the Institute's focus area.

Many SMEs experience common challenges that hamper their performance. They face problems with lagging productivity and business practices, missed opportunities in emerging technologies, unrealized growth potential, and leadership and succession issues. They report significant challenges in ongoing continuous improvement, identifying growth opportunities, and product innovation/development for manufacturability. As significant sources of innovation in highly sophisticated supply chains, they face increasing demands from their original equipment manufacturer (OEM) customers for improved performance, quality, consistency, and continued innovation.

An IMI may help SMEs address challenges in a myriad of ways. IMIs can provide information about technology trends and access to cutting edge technologies that assist with process innovations and development. They can provide shared facilities and access to specialized equipment that can accelerate product design, prototyping, and testing. IMIs can provide technical advice and assistance to SMEs that do not have specialized expertise on staff. An IMI should also facilitate the creation of new, start-up companies to commercialize research and development results.

SMEs interested in a broad range of services and an ongoing relationship might participate in a tiered membership structure that would minimize barriers to entry and encourage the membership of SMEs in the Institute.

## 3.6. IMI Governance

Each Institute should have substantial autonomy from its partner organizations and institutions and should have an independent fiduciary Board of Directors predominantly composed of industry representatives. An Institute leader such as an Executive Director should be in charge of day-to-day operations. The three key stakeholders of NNMI (industry, academia, and government) will need to have their interests preserved in a joint governance model.

Each proposal will outline the methods by which decisions will be made, including those decisions related to operations, membership, intellectual property, capital investments, project selection, funding allocation, and progress toward sustainability. Draft Institute membership and governance agreements should be included within proposals.

## 3.7. IMI Selection Process

The IMI selection process will be managed by the interagency AMNPO. Participating Federal agencies currently include DOC/NIST, DOD, DOE, ED, NASA, and NSF. The inter-agency AMNPO team will be responsible for managing an open, competitive selection process and for executing the award process.

Solicitations for Institute proposals may be staged, and the design and number of solicitations will depend on the availability and timing of funds. Proposals received in response to the solicitation(s) will be evaluated competitively by a review team.

The review team will include members of the AMNPO, agency partners, and other experts. The merit-based selection process may include pre-proposals, site visits, and economic and business plan analyses. Support for proposals may be offered by the AMNPO through planning and/or road mapping grants and workshops. This broad, deliberative review process can best balance the most essential U.S. industrial needs and promising opportunities and support the goals of enhancing American industrial competitiveness on regional, national and global bases.

Detailed selection criteria will be contained in the IMI solicitation(s). It is expected that criteria will include:

1) The focus on a critical national need or opportunity for U.S. manufacturing
2) Proposed activities targeting the transition of early stage manufacturing research and technology to commercial application or product
3) The proposed Institute plan to achieve significant impact in manufacturing technology development and scale up to commercialization, anticipating widespread adoption and links to job creation and broad economic impacts
4) The proposed Institute resources (personnel, facilities, and participating entities) supporting the plan
5) The level of co-investment from non-Federal entities, and the strength of the plan for sustain-ability beyond the initial Federal funding
6) The adequacy of the financial plan
7) The level of engagement with SMEs
8) The suitability and anticipated utilization of shared facilities
9) The level of involvement and expertise in education and workforce development
10) The adequacy of the governance and oversight model, including the degree of industry involvement and openness to relevant new participants

11) The ability of the Institute to advance American domestic manufacturing

It is likely that cooperative agreements will be the primary funding mechanism for NNMI Institute awards, although other types of grants and contracts may also be used. Continuing Federal support will be contingent on co-investment by businesses and other non-Federal entities, and on progress toward sustainable operations.

## 3.8. Disbursement of Federal Funds

A schedule of disbursement of funds will be negotiated for each IMI depending on the nature of the Institute and proposed projects. It is anticipated that for most Institutes, the level of Federal funding will be larger at the start, and will become progressively lower as the Institute becomes self-sufficient, as shown in Figure 1.

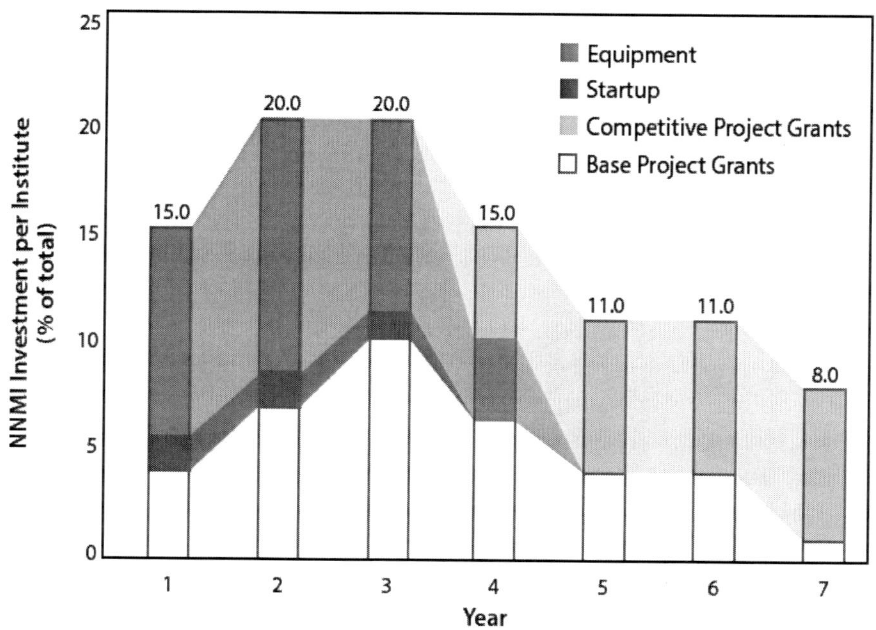


Figure 1. Notional Federal funding profile for an IMI.

Figure 1 shows the envisioned Federal investment in an exemplar, capital equipment-intensive Institute, and is not a funding profile that will be required of all Institutes. Federal funding is shown during the first year for Equipment, Startup, and an initial slate of Base Projects. As the Institute matures, it begins to absorb more of its own operational costs. During year four, Federal funding of projects begins to depend on the Competitive Project Grants. The pool of funds available for the competitive project grants is approximately 25% of the total funding provided to the Institutes. The competitive project grants will be relatively small in number, totaling 2 or 3 per Institute beginning in year 4, and relatively large in scale, each around $2–3 million. The competitive project grants will encourage increasing partner co-investment as the Institute matures.

Continuation of Federal funding will be contingent on the completion of annual reports, and on the successful completion of a gate review process examining both quantitative and qualitative metrics of performances including: co-investment, membership, facility utilization, project portfolio, success stories, technology commercialization, or other benchmarks that assess Institute performance and impact. It is anticipated that a gate review will occur at least at year three after initial funding.

Not shown in Figure 1 is the co-investment funding provided by non-Federal sources. The non-Federal support is anticipated to be large at the time of award and primarily comprised of in-kind items such as equipment and buildings. Over time, the funding will shift to project, member fees, user fees, licensing and other sources.

## 4.0. National Network of Manufacturing Innovation Institutes

To ensure the IMI's fully leverage their potential to advance American domestic manufacturing, the National Network for Manufacturing Innovation will support and expand the impacts of individual Institutes, develop cross-cutting metrics and methods for evaluating impact, and promulgate best practices and standards. Each Institute will participate in the broader National Network. The purpose and functions of the Network are outlined below.

## 4.1. Inter-Institute Collaboration

To the extent possible, the Institutes should work collaboratively, sharing resources, best practices, and research and development results. They should transparently share funding and membership models, annual reports, and projections. The Institutes are not directly competitive, as IMIs will have diverse goals, but rather they will share the broad mission of improving U.S. manufacturing competitiveness. To support these goals, the NNMI will organize a Network Leadership Council composed of representatives of the Institutes, Federal agencies, and other appropriate entities. The Network Leadership Council will actively look for opportunities to leverage existing resources between Institutes.

## 4.2. Common IMI Policies

Efficient operation of the Network will be facilitated through common policies. Common policies facilitate interaction with SMEs, promote collaboration and movement within the Network, and allow IMIs to share services such human resource management. While recognizing the differing needs of various manufacturing sectors, clusters, and ecosystems, the Network will strive, as far as is practical, to maintain common policies with regard to intellectual property, contract research, operations, accountability, and marketing and branding.

## 4.3. Manufacturing Portal Participation

While each Institute may have its own web presence, they will also participate in and link to the AMNPO hosted “Manufacturing Portal” (www.manufacturing.gov). As the network and funded Institutes emerge, this portal will serve to direct interested parties to the resources and capabilities embodied within the overall Network. Content will include information about the focus of each Institute, structure, governance, contacts, annual reports, news, success stories, intellectual property available for licensing, member information, and more.

## 5.0. RATIONALE FOR THE INVESTMENT

In his 2012 State of the Union address, President Obama laid out his "blueprint for an economy that's built to last—an economy built on American manufacturing, American energy, skills for American workers, and a renewal of American values."

"This blueprint," he said, "begins with American manufacturing." [22]

The President chose this starting point for good reason. Numerous recent reports have documented the critical role of U.S. manufacturing to innovation [11], jobs [23, 25], the economy [25], exports [25, 26], and national security [27]. The President has initiated a set of actions designed to make our manufacturing sector more competitive and to encourage more investment in the United States. These actions encompass sound tax policies, enforcement of trade laws, and investments in innovation, advanced technology, education, and infrastructure.

According to the non-partisan Council on Competitiveness, "U.S. manufacturing is more important than ever." [28] While not dismissing serious challenges posed by low-wage international competitors and rivals that are fast advancing in technological capabilities, the Council maintains that significant opportunities to increase production and grow exports exist for U.S. manufacturers. "The digital, biotechnology, and nanotechnology revolutions," it reports, "are unleashing vast opportunities for innovation and manufacturing." [28]

Moving to put his blueprint into action, the President has launched the NNMI program to advance American domestic manufacturing by strengthening the innovation, competitiveness, and job-creating power of U.S. manufacturing. The Network will help to address an inconsistency in U.S. economic and innovation policies. The Federal Government investment in public R&D reached more than $133 billion in FY 2009 [24], and a tax credit exists for industry-funded R&D. Yet, these measures are not matched by corresponding, strategically designed and implemented efforts and incentives to encourage domestic manufacturing of products ultimately arising from U.S. discoveries and inventions [6]. Recognizing the severity of the problem, the administration has recently taken additional action to accelerate technology transfer and commercialization of Federal research [29].

As documented by the National Science and Technology Council, "A gap exists between R&D activities and the deployment of technological innovations in domestic production of goods," [8] contributing significantly, for example, to the growing trade deficit in advanced technology products. In

2011, the United States ran a $99 billion deficit in trade of advanced technology products, accounting for 17 percent of the total U.S. trade deficit [30]. The United States has lost 687,000 high-technology manufacturing jobs since 2000 [24], when the nation posted a $5 billion trade surplus in advanced technology products.

Manufacturing plays a disproportionately large—and valuable—role with respect to the nation's innovation capacity, generating $1.35 in economic activity for every $1 of sector output [4]. Compared to all other sectors, manufacturing has the largest multiplier [31]. This figure can be contrasted to the service sectors (including financial services) that generate between $0.55 and $0.66 in additional economic activity for every $1 of sector output. Manufacturing accounts for about 12 percent of the Nation's gross domestic product, but accounts for 70 percent of domestic industry R&D spending and employs 60 percent of industry's scientists and engineers. Thus, manufacturing remains the essential core of the U.S. economy's innovation infrastructure.

Accelerating innovation and implementation of advanced manufacturing capabilities requires bridging a number of gaps in the present U.S. innovation system. Time horizons for investment payback in pre-commercial manufacturing innovation often exceed investor expectations, and technical and commercial risks are greater than they are for investments aimed at incremental improvements in existing products and services.

The commonly-held belief that knowledge spillovers reduce returns from privately-funded innovation is a related obstacle to patient, sustained private-sector investment in developing promising technologies all the way through to manufacturing and commercial market introduction. Because of the perceived so-called "free rider" problem, no single company will typically take on the risk and devote the resources needed to build the full infrastructure of underpinning manufacturing capabilities and complementary resources that would benefit an entire industry and even groups of industries. This is an area where the Federal Government can play a role to lessen that risk and help to catalyze building of an advanced-manufacturing infrastructure.

Historically and even today, the United States has excelled at basic science and invention. But the commercial and economic rewards that can sprout and grow from these important early-stage accomplishments are realized in the stages following initial discovery—especially at the points of manufacturing scale-up and commercialization. However, many technologies rooted in U.S. research fail to mature to full scale-up and commercialization in domestic factories.

## 5.1. NNMI in the Context of 21st Century Global Competition

Other nations also recognize the strong links between manufacturing, innovation, and prosperity. Many have established programs to strengthen the links. Currently, Germany, Korea, and Japan each have more R&D-intensive manufacturing sectors than the United States.

As technologies and products become ever-more complex and their product-development cycles shrink, successfully mastering all the stages from laboratory to marketplace requires contributions from a large network of organizations—from suppliers of equipment, parts, and services; to schools, colleges, and training programs; to utilities and other infrastructure systems. As global competition to manufacture and sell high-value-added products intensifies, the capabilities and performance of these innovation ecosystems must improve.

Many U.S. competitors also recognize the importance of these location-based linkages, and governments of these nations are investing significantly to strengthen and expand their clusters of innovation and advanced manufacturing capabilities.

In the fast-paced, highly competitive global economy, national competitiveness continues to have strong local underpinnings. Pisano and Shih define the competitiveness of a country as "the advantage workers and organizations located in one place—a local commons—enjoy in the production of specific goods or services over workers and organizations located elsewhere [32].

For the United States and other nations, national competitiveness is largely the composite result of the performance (innovation, manufacturing, and other dimensions of competitiveness) of many regions. As a result of many factors, the competiveness of a number of U.S. regions has waned over the last decade or more. One serious result with long-term economic consequences is the hollowing out of the so-called regional *industrial commons*—"the set of manufacturing and technical capabilities that support innovation across a broad range of industries" [33].

A concept closely related to industrial commons is *infrastructure*. The modern infrastructure needed to support advanced, high-tech industries has many dimensions—transportation, communication, intellectual, workforce skills, capital availability, and others. The United States has many infrastructural strengths that can be leveraged to maximize the competitive and innovation performances of U.S. industries, among them "a world-class university system, strong intellectual property protection, sophisticated managerial talent, ready access to capital, and a huge domestic market" [34].

Also included among these competitive assets is an array of private and public (Federal, State, local) programs with innovation-focused (or related) missions. Examples are consortia, university research parks, university-industry research centers, incubators, national laboratories, manufacturing extension services, skill certification programs for workers, and State and regional economic development programs. By themselves, some of these efforts have yielded commendable results.

In addition, efforts in the United States and other nations to spur innovation, build advanced manufacturing capabilities, and spawn new industries and jobs have been focused on geographic regions. Successful cluster-based approaches yield benefits that foster long-term competitiveness, including a skilled workforce, inter-firm linkages, and knowledge spillovers—all of which contribute significantly to innovation as integral components of regional industrial commons [6, 33].

While the United States has a variety of programs and activities that aim to strengthen regional innovation clusters or to catalyze the emergence and growth of new ones, the overall effort has been described as fragmented or uncoordinated [34, 35]. The lack of coordination and programmatic linkages means that the individual programs and activities (private, Federal, State, and local) may not be as effective as they could be in promoting the development, application, and commercialization of new product and process technologies, as well as new manufacturing capabilities.

To maximize chances for success, the National Research Council's report [6] advises that regional innovation clusters need to:

1) Leverage local strengths,
2) Encourage self-organization,
3) Pool resources,
4) Share risks,
5) Grow a trained workforce,
6) Connect with local universities and laboratories,
7) Provide long-term commitment,
8) Provide incentives, and
9) Monitor and measure industry needs

## 5.2. Overview of Existing Relevant Programs

While the Federal Government has a number of highly effective programs in its R&D portfolio for advanced manufacturing, there are no current Federal programs that cover NNMI attributes to the extent needed to significantly influence the nation's competitiveness. A balanced portfolio of targeted programs (basic research through applied scale-up) is required for a complete national response to the challenges and opportunities in advanced manufacturing. As a means of contrast, selected programs are described below and compared to the NNMI Institute vision:

### *5.2.1. Research Centers*

Research centers typically focus on basic research, but occasionally also extend to fund applied research and technology proof-of-concept activities. Such centers are affiliated with one or more universities. Examples include:

1) NSF Industry/University Cooperative Research Centers that leverage a modest NSF investment with funding by industry members, other agencies, and other organizations to support a single- or multi-university center. The centers focus on early-stage precompetitive research.
2) NSF Engineering Research Centers are hosted at universities and promote partnerships among researchers in different disciplines and between industry and universities. ERCs are targeted at advancing transformational engineering systems, conduct research from the fundamental stage to proof-of-concept test beds, create education programs from pre-college through postdoctoral studies, and are hosted by universities. The goal is to create a culture in engineering research and education that links discovery to technological innovation through transformational fundamental and engineered systems research.
3) NSF's Materials Research Science and Engineering Centers (MRSECs) support researchers to engage in fundamental interdisciplinary research to solve complex materials problems that are intellectually challenging and important to society. Discoveries resulting from this research are sometimes leveraged into start-up companies impacting industries across the breadth of materials science. Additionally, centers are encouraged to engage in external collaborations, which can include scientists from industry.

In addition, Federally Funded Research and Development Centers (FFRDCs)—although not focused specifically on manufacturing R&D—conduct basic and applied research and support specialized R&D resources and capabilities.

### *5.2.2. Research and Innovation Hubs*

Hubs emphasize large and diverse teams of researchers focused on a specific high priority goal. For example, the DOE Energy Innovation Hubs are multi-disciplinary, multi-institutional, highly-collaborative teams of scientists and engineers working over a longer time frame to advance highly promising areas of energy science and technology from their early stages of research to the point where the risk level will be low enough for industry to begin to move them into development for manufacturing scale-up and the marketplace. DOE Hubs are led by top researchers with the knowledge, resources, and authority to nimbly guide efforts, seizing new opportunities or closing off unproductive lines of research.

### *5.2.3. Manufacturing Demonstration Facilities*

MDFs help members of the manufacturing community to develop and demonstrate new manufacturing technologies. This can help manufacturers understand full-scale costs and implementation requirements, allowing them to create and justify their business case when seeking further private capital investment to bring processes and products to market. MDFs access to otherwise cost-prohibitive tools and resources, and are focused on demonstration of processes and production technologies at industrially relevant demonstration scale. DARPA and DOE are sponsors of MDFs.

### *5.2.4. DOD Manufacturing Technology (ManTech) Program*

The DOD ManTech Program develops advanced manufacturing technologies and processes for the affordable, timely production and sustainment of defense systems. In close partnership with industry, the program impacts all phases of system development, acquisition, and sustainment by developing, maturing, and transitioning key advanced manufacturing technologies. Investments are focused on those technologies that have the greatest defense benefit and are balanced to support transi tion of emerging technologies, improvements to existing production enterprises, and strengthening the U.S. industrial base. ManTech has a long history of delivering critical and "game changing" advanced manufacturing technologies and processes, such as numerically controlled machines, carbon fiber composites, microelectronics fabrication, advanced radars, laser-guided munitions, lean production methods, advanced optics, and advanced soldier body armor. Many defense manufacturing technologies have spun-off into commercial markets and resulted in large economic advances for the United States. For example, the widespread commercialization of numerically controlled machine tools has been attributed to initial ManTech sponsorship and development.

### *5.2.5. Centers of Excellence*

Centers of Excellence are another model used in a number of Federal programs. These centers are typically led by a not-for-profit or university and concentrate basic or applied research into a specific area in order to achieve a critical mass of talent and resources. DOD's Navy ManTech Program utilizes a Center of Excellence model to develop technologies and processes for the affordable production and sustainment of naval systems. Centers of Excellence have existed for Joining, Composites, Electronics, and Benchmarking Best Practices. DOE's Fuel Cell Technology Program previously sponsored four Hydrogen Storage Centers of Excellence in order to accelerate optimal materials for chemical and metal hydride hydrogen storage.

### *5.2.6. Deployment and Delivery Programs*

Deployment and Delivery Programs provide hands-on technical assistance to companies interested in implementing lessons from classroom training or industry best practice into their business growth strategies and manufacturing operations. This includes employing new technologies to meet changing customer demands or support new product development, deploying innovative production methods, or other steps to enhance their capabilities and competitive position. For example, the DOC NIST Hollings Manufacturing Extension Partnership is a national network based on a Federal-State-local partnership providing direct services to nearly 10,000 SMEs annually through 400 centers, field offices, and partner organizations across the U.S. MEPs provide fee-based technical assistance to manufacturers to enable them to reduce costs, improve productivity, and pursue growth opportunities through innovation.

### *5.2.7. Federal Support for Advanced Manufacturing Workforce Development*

There are several Federal programs and activities in support of developing and nurturing a workforce skilled in advanced manufacturing. For example:

1) DARPA's Manufacturing Experimentation and Outreach (MENTOR) effort engages high school students in collaborative, distributed manufacturing and design.
2) State directors of career and technical education, along with the Department of Education, are promoting the National Career Clusters Framework, which supports quality career and technical education programs through learning and comprehensive programs of study. Manufacturing is one of 16 Career Clusters, guiding development of programs of study in manufacturing that bridge secondary and postsecondary curricula.
3) The Department of Labor's Employment and Training Administration (ETA) supports the development of a skilled manufacturing workforce through the Registered Apprenticeship program, the Workforce Investment Act programs, and the manufacturing competency model. This employer-validated model outlines the skills necessary to pursue a successful career in the manufacturing industry. ETA's competitively awarded grants also support efforts to accelerate education and training programs for high-skilled occupations in the advanced manufacturing field. The Trade Adjustment Assistance Community College and Career Training Grant Program, for example, catalyzes partnerships between community colleges, the public workforce system and employers to develop innovative educational training models in advanced manufacturing, among other critical industry sectors. By raising technical skill levels, American workers can obtain the industry-recognized credentials required to enter and advance along career pathways in the manufacturing sector.

## 5.3. Differences between Existing Programs and the NNMI Vision

The notable distinctions between NNMI Institutes and existing Federal programs include:

1) **Federal investment level:** The I/UCRC and ERC funding levels are an order of magnitude lower per center than is planned for Institutes. Such centers have a structure typically aligned with university policies and requirements, and do not have the Institute's industry focus. Workforce development is addressed in ERCs, but community

colleges are not an integral part of an ERC. Leadership originates with research universities, while with Institutes, leadership is provided by industry, as well as government, academic, public and private laboratories, and any other affected parties that are stakeholders within that IMI. Research projects for both of the NSF center programs described above target the early stage fundamental research to proof-of-concept.

2) **Focus on prototyping and scale-up:** The focuses of many of the existing Federal programs and the proposed Institutes are fundamentally different. With existing Federal programs, the investment is primarily in basic and applied early stage research without a specific focus on manufacturability and manufacturing processes and technologies. Existing industry investment is predominantly in late stage research and demonstration and incremental process development. The region between these two investment areas is recognized as an underfunded and critical area. One of the fundamental concepts central to the NNMI is that a government-industry-academia *partnership* will leverage the resources of everyone to address gaps in innovation in the manufacturing space, which includes both developing manufacturing technologies and designing/modifying products to enable their manufacture.
3) **MEP centers** were intended to pull applied research, technologies, and innovations from universities, Federal labs, and other sources to support client SME growth initiatives. Historically, MEP centers have focused on SME process improvements and quality issues. In recent years, the program has sought to build on that expertise and SME client relationships to accelerate the rate of adoption of new technologies by clients. MEP connects technology opportunities and SMEs to move new product opportunities into production and into the market faster and with greater returns on investment. This relatively nascent MEP initiative would complement NNMI directions.
4) **Industry-driven:** In the Institutes, industry will have a strong role in the identification of the technology area and research agenda through their co-investment. NAMII's technology focus on additive manufacturing, in contrast, grew largely out of Federal agency needs (although the topic is widely supported by industry as a critical focus area). In addition, the industry-wide cross-cutting focus of NNMI

Institutes will often span the research interests of many industry partners, academic partners, and multiple Federal agencies.
5) **Significant industry co-investment:** With NNMI, the Federal funding will be heavily leveraged by Institute partners'direct co-investment at a sufficiently large scale to affect regional economies or entire industry sectors. This will increase the impact of Federal funding. It will force the Institute to have a pragmatic focus on industrially relevant technologies, and also will empower the partners as true stakeholders in the Institute's success.
6) **Network attributes:** The NNMI program includes a vision for the Institutes to interact through a National Network.

# 6.0. National Additive Manufacturing Innovation Institute— The Pilot Institute

In March 2012, when President Obama unveiled his proposal to build the NNMI, he also initiated steps to jumpstart the NNMI with a Pilot Institute. The Pilot Institute would serve as a proof-of-concept for the NNMI Institutes. A collaborative interagency team was convened and determined that the topic of Additive Manufacturing would garner the most benefit for the defense, energy, space, and commercial sectors of the nation, and should be the focus of the Pilot Institute. Additive Manufacturing, also commonly known as 3D printing, is an emerging and evolving collection of manufacturing processes that build metal, plastic, or ceramic parts using layer-by-layer build-up techniques, precisely placing material as directed, based on a software representation of the three-dimensional part geometry.

The competition for the Pilot Institute was launched through a DOD Broad Agency Announcement (BAA) in May 2012. Proposals for the Pilot Institute were due on June 16, and the award to the National Additive Manufacturing Innovation Institute (NAMII) was announced on August 16, 2012 [36]. NAMII is headquartered in Youngstown, Ohio, and it is comprised of a broad coalition of more than 80 companies, 9 research universities, 6 community colleges, and 18 not-for-profit institutions.

While NAMII is the Pilot Institute, it differs from the proposed IMIs in several important ways. The focus area was pre-defined based on agency needs, funding was obtained from multiple agencies, and the partner agencies set the capitalization at a significantly lower level than is envisioned for an IMI.

Proposal evaluations were conducted by an interagency advisory council of technical experts from the DOC, DOD, DOE, NASA, and NSF. Based on the evaluation process stated within the BAA, the advisory council selected the National Center for Defense Manufacturing and Machining (NCDMM) to manage the Pilot Institute. This public-private partnership between NCDMM and the Government was awarded as a cooperative agreement using $30 million of Federal funding and an additional $39 million provided as cost share, mostly from industry and the states of Ohio, Pennsylvania, and West Virginia.

This long-term, public-private partnership between State and local governments, industry, and academia (including research universities and community colleges) is enabled by the Federal Government and, as such, DOD and other partnering agencies used cooperative agreements (rather than grants or contracts) to fund the government portion. A governance board along with an executive committee and technical advisory board comprised of industry, academia, and government representatives oversee the Institute's activities. Using a cooperative agreement also allows for substantial involvement by the Federal Government, and is consistent with the broad interest in additive manufacturing across the multiple agencies that collaborated to launch the pilot initiative.

In summary, NAMII was formed in parallel with the ongoing effort to architect the broader NNMI and hence is technically not part of a yet-to-be-formed network. However, it represents a critically important seeding of the concept and its construction and early activities are helping to inform the design and development of the full NNMI enterprise.

## 7.0. Crowd Sourcing the Design of the NNMI

To develop a knowledge base regarding the President's proposed NNMI, and to assist in the NNMI design, the AMNPO embarked on a nationwide strategy to solicit input from stakeholders from industry, academia, State and regional governments, economic development authorities, industry associations and consortia, private citizens, and other interested parties.

## 7.1. NNMI Request for Information

On May 4, 2012, the AMNPO, through NIST, published a Request for Information (RFI) in the Federal Register inviting interested parties to provide public comment on the new proposed public-private NNMI program [37]. Comments in response to the RFI were accepted through October 25, 2012. The RFI sought open input on the NNMI and specific input on 21 questions in four categories related to the structure and operations of the individual Institutes and the NNMI. The topics and questions were:

***Technologies with Broad Impact***

1) What criteria should be used to select technology focus areas?
2) What technology focus areas that meet these criteria would you be willing to co-invest in?
3) What measures could demonstrate that Institute technology activities assist U.S. manufacturing?
4) What measures could assess the performance and impact of Institutes?

***Institute Structure and Governance***

1) What business models would be effective for the Institutes to manage business decisions?
2) What governance models would be effective for the Institutes to manage governance decisions?
3) What membership and participation structure would be effective for the Institutes, such as financial and intellectual property obligations, access and licensing?
4) How should a network of Institutes optimally operate?
5) What measures could assess effectiveness of Network structure and governance?

***Strategies for Sustainable Institute Operations***

1) How should initial funding co-investments of the Federal Government and others be organized by types and proportions?
2) What arrangements for co-investment proportions and types could help an Institute become self-sustaining?
3) What measures could assess progress of an Institute towards being self-sustaining?

4) What actions or conditions could improve how Institute operations support domestic manufacturing facilities while maintaining consistency with our international obligations?
5) How should Institutes engage other manufacturing related programs and networks?
6) How should Institutes interact with State and local economic development authorities?
7) What measures could assess Institute contributions to long-term national security and competitiveness?

***Education and Workforce Development***

1) How could Institutes support advanced manufacturing workforce development at all educational levels?
2) How could Institutes ensure that advanced manufacturing workforce development activities address industry needs?
3) How could Institutes and the NNMI leverage and complement other education and workforce development programs?
4) What measures could assess Institute performance and impact on education and workforce development?
5) How might Institutes integrate research and development activities and education to best prepare the current and future workforce?

In total, the AMNPO received seventy-eight (78) separate RFI responses from industry, academia, economic development, State and regional authorities, national laboratories, and private citizens, representing the viewpoints of nearly more than 100 separate entities [38].

## 7.2. NNMI Regional Workshops

In addition to the RFI as a crowd sourcing tool, the AMNPO held four regional workshops as part of its strategy for soliciting input. The four *Designing for Impact* regional workshops attracted approximately 875 attendees from across the country. Participants included leaders and representatives from industry; academia; economic development organizations; State, local and Federal Government; and other organizations (see Figure 2). The four workshops were:

1) **Designing for Impact I:** April 25, 2012, at Rensselaer Polytechnic Institute in Troy, New York
2) **Designing for Impact II:** July 9, 2012 at Cuyahoga Community College outside Cleveland, Ohio
3) **Designing for Impact III:** September 27, 2012 at the Arnold and Mabel Beckman Center of the National Academies of Sciences and Engineering in Irvine, California
4) **Designing for Impact IV:** October 18, 2012 at the Millennium Harvest House in Boulder, Colorado.

Reports [39] from all four workshops summarizing the individual recommendations and comments of workshop attendees have been used to assist in the design of the NNMI.

Through these outreach efforts, the Federal agency partners explored a wide-ranging set of issues central to the design of the NNMI, including technology focus areas, institutional design and governance issues, and management of the network as a whole to amplify the impact of individual Institutes.

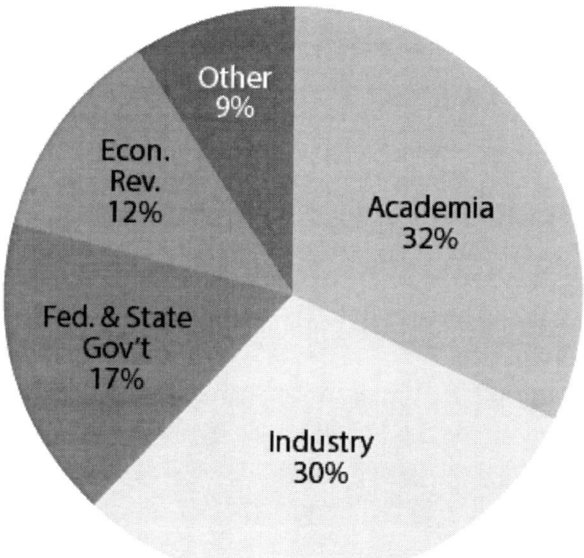


Figure 2. *Designing for Impact I-IV* Workshop Participation by Sector.

## REFERENCES

[1] U.S. Bureau of Economic Analysis, “1998-2011 Summary Make Annual I-O Table Before Redefinitions," http://www.bea.gov/industry /io_ annual.htm

[2] Bureau of Labor Statistics, Quarterly Census of Employment and Wages (Establishment Size Data Available from QCEW), http://www.bls.gov /cew/cewsize.htm

[3] U.S. Census Bureau, Trade in Goods with Advance Technology Products, 2010 and 2011.

[4] Bureau of Economic Analysis, Industry-by-Industry Total Requirements after Redefinitions (Producer Price Indexes), 2010, www.bea.gov /iTable/index_industry.cfm

[5] World Economic Forum, The Future of Manufacturing: Opportunities to Drive Economic Growth, April 2012.

[6] Rising to the Challenge: U.S. Innovation Policy for Global Economy, National Research Council, 2012, http://www.nap.edu/catalog.php? record_id=13386#toc

[7] Capturing Domestic Competitive Advantage in Advanced Manufacturing, AMP Steering Committee Report, President’s Council of Advisors on Science and Technology July 2012, http://www.white house.gov/sites/default/files/microsites/ostp/pcast_amp_steering_commi ttee_ report_final_july_17_2012.pdf

[8] A National Strategic Plan for Advanced Manufacturing, February 2012, http://www.whitehouse. gov/sites/default/files/microsites/ostp/iam_advancedmanufacturing_strat egicplan_2012.pdf

[9] International Benchmarking of Countries’ Policies and Programs Supporting SME Manufacturers,The Information Technology and Innovation Foundation, September 2011, http://itif.org/publications /international-benchmarking-countries%E2%80%99-policies-and-programs-supporting-sme-manufacturer

[10] Manufacturing Growth: Advanced Manufacturing and the Future of the American Economy, Breakthrough Institute-Third Way-Bernard L. Schwartz Initiative on American Economic Policy, October 2011, http://thebreakthrough.org/

[11] Report to the President on Ensuring American Leadership in Advanced Manufacturing, June 2011, http://www.whitehouse.gov/sites/default/files /microsites/ostp/pcast-advanced- manufacturing-june2011.pdf

[12] The Federal Role in Supporting Urban Manufacturing, What Works Collaborative April 2011, http://www.areadevelopment.com/article_pdf/id61372_04_urban_manufacturing_mistry_byron. pdf
[13] The Case for a National Manufacturing Strategy, The Information Technology & Innovation Foundation, April 2011, http://www2.itif.org /2011-national-manufacturing-strategy.pdf
[14] Manufacturing Resurgence, A Must for US Prosperity, National Association of Manufacturers August 2010, http://www.nam.org /~/media /13A10411852D446F84977230D8DD5B8D.ashx
[15] Manufacturing Mandate, Association of Manufacturing Technology 2010, http://www.amtonline.org/images/userContent/Manufacturing Mandate_Book_Ver04.pdf
[16] The New 'Cluster Moment': How Regional Innovation Clusters Can Foster the Next Economy, Brookings Metropolitan Policy Program 2010, http://www.wedc.wa.gov/Download%20files/2010.09-Cluster Moment-Brookings.pdf
[17] Catalyzing American Ingenuity: The Role of Government in Energy Innovation, American Energy Innovation Council Sept. 2011, http://bipartisanpolicy.org/sites/default/files/AEIC%20Report.pdf
[18] President Obama to Announce New Efforts to Support Manufacturing Innovation, Encourage Insourcing, http://www.whitehouse.gov/the-press-office/2012/03/09/president-obama-announce-new-efforts-support-manufacturing-innovation-en
[19] National Science Board, Science and Engineering Indicators 2010, http://www.nsf.gov/statistics/seind10/pdf/c04.pdf
[20] http://www.whitehouse.gov/sites/default/files/omb/budget/fy2013/assets/topics.pdf
[21] Technology Readiness Assessment Deskbook, July 2009, http://www.skatelescope.org/public/2011-11-18_WBS-SOW_Development_Reference_Documents/ DoD_TRA_July_2009_Read_Version.pdf
[22] Remarks by the President in State of the Union Address, January 24, 2012, http://www.whitehouse.gov/the-press-office/2012/01/24/remarks-president-state-union-address
[23] Bureau of Labor Statistics, 2011 Employer Costs for Employee Compensation, Table 6, http://www.bls.gov/news.release/archives /ecec_06082011.pdf
[24] National Science Board, Science and Engineering Indicators 2012, http://www.nsf.gov/statistics/seind12/
[25] Bureau of Economic Analysis, http://www.bea.gov/industry/index.htm

[26] U.S. Census Bureau, U.S. Bureau of Economic Analysis, NEWS, September 11, 2012, http://www.bea.gov/newsreleases/international /trade/2012/pdf/trad0712.pdf

[27] A National Strategic Plan for Advanced Manufacturing, February 2012, http://www.whitehouse.gov/sites/default/files/microsites/ostp/iam_advan cedmanufacturing_ strategicplan_2012.pdf

[28] MAKE An American Manufacturing Movement, December 2011, http://www.compete.org/images/uploads/File/PDF%20Files/USMCI_M ake.pdf

[29] "Accelerating Technology Transfer and Commercialization of Federal Research in Support of High Growth Businesses," Presidential Memorandum October 28, 2011, http://www.whitehouse.gov/the-press-office/2011/10/28/ presidential-memorandum-accelerating-technology-transfer-and-commerciali

[30] U.S. Census Bureau, "Trade in Goods with Advance Technology Products," 2011, https://www.census.gov/foreign-trade/balance/c0007. html#2011

[31] The Manufacturing Institute, The Facts about Modern Manufacturing, 8th Edition, 2009, www.nist.gov/mep/upload/FINAL_NAM_ REPORT_PAGES.pdf

[32] Producing Prosperity, Gary P. Pisano and Willy C. Shih, Harvard Business Review Press, 2012.

[33] "Choosing the United States," Michael E. Porter and Jan W. Rivkin, Harvard Business Review, March 2012.

[34] National Research Council, Growing Innovation Clusters for American Prosperity, The National Academies Press, 2011.

[35] "Clusters and Economic Policy: Aligning Public Policy with the New Economics of Competition," Michael Porter, ISC White Paper, Harvard Business School, November 2007.

[36] National Additive Manufacturing Innovation Institute, http://namii.org/

[37] Federal Register /Vol. 77, No. 87 / Friday, May 4, 2012 /Notices, http://www.gpo.gov/fdsys/pkg/FR-2012-05-04/pdf/2012-10809.pdf

[38] RFI Responses & Workshop Reports, http://manufacturing.gov /rfi_responses.html

[39] Publications and Resources, http://www.manufacturing.gov/pubs_ resources.html

# Appendix A. NNMI Public Workshop and RFI Participants

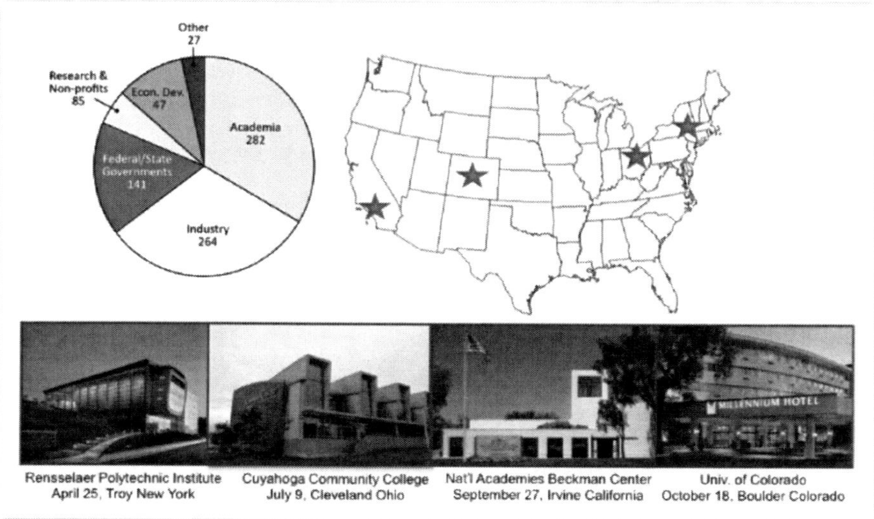


## Academia (282)

Alfred University
Arizona State University
Association of Public and Land-grant Universities
Case Western Reserve University
City University of New York
Columbia University
Iowa State University
Kansas State University
Kent State University
Lorain County Community College
Macomb Community College & Auto Communities Consortium
National Institute for Pharmaceutical Technology and Education
Pennsylvania State University - Electro-Optics Center
Purdue University
The University of Akron
University at Albany – CNSE

University at Buffalo
University of California Davis – College of Engineering
University of California, Berkeley - College of Engineering
University of Illinois
University of North Carolina at Charlotte
University of Rochester – Center for Emerging and Innovative Sciences
University of Southern California – Information Sciences Institute
University of Texas at Arlington Research Institute
Youngstown State University
Abbaschian, Reza, University of California, Riverside
Abt, Steve, Colorado State University
Adams, George, Purdue University
Aidun, Daryush, Clarkson University
Anaya, Jose, El Camino College
Anderson, Dennis, North Dakota State University
Anderson, Sheri, North Dakota State University
Anthony, Brian W., Massachusetts Institute of Technology
Apanius, Matt, Lorain County Community College
Apelian, Diran, Worcester Polytechnic Institute
Baecker, Charlie, Beall Center for Innovation and Entrepreneurship
Baeslack, William, Case Western Reserve University
Balaji, A.K., The University of Utah
Banerjee, Amarnath, Texas A&M University
Barako, Tristan, Brown University
Barnak, Gary, Saddleback College
Batchelor, Ann, CSU Ventures
Beaman, Joseph, University of Texas
Beaudoin, Steve, Purdue University
Benson, Abby, University of Colorado Offce of Government Relations
Bergeson, Eric, Clemson University
Berman, Fran, Rensselaer Polytechnic Institute
Beyer, Christiane, California State University, Long Beach
Birken, Ralf, Northeastern University
Black, Randy, University of California, Irvine
Bohlmann, Brad, University of Minnesota
Boretz, Mitch, University of California, Riverside
Bowman, Keith, Illinois Institute of Technology
Braggins, Alan, San Bernardino CCD
Brasche, Lisa, Iowa State CNDE

Bridger, Amy, Penn State Erie
Brookstein, David, Philadelphia University
Busnaina, Ahmed, Northeastern University
Cakmak, Miko, University of Akron
Camelio, Jaime, Virginia Polytechnic Institute and State University
Camp, Lisa, Case Western Reserve University
Cantello, Craig, Edison Tech Center
Canton, Giulia, University of California, Irvine
Carron, Alice, Navajo Technical College
Cetinkaya, Cetin, Clarkson University
Chen, Shaochen, Univ of California, San Diego
Chen, Julie, University of Massachusetts, Lowell
Chen, Yong, University of Southern California
Chiumento, Laurie, University of Rochester
Church, Roy, Lorain County Community College
Chyu, Ming King, University of Pittsburgh
Cormier, Denis, Rochester Institute of Technology
Corson, Paul, Lorain County Community College
Damoulakis, John, Information Sciences Institute
Davis, Jim, University of California, Los Angeles
Dean, David, Case Western Reserve University
DeSilva, Brett, Saddleback College
Diaz, Bob, College of Southern Nevada
Donnellan, Thomas, Applied Research Laboratory at the Pennsylvania State University
Dordick, Jonathan, Rensselaer Polytechnic Institute
Dory, Craig, Rensselaer Polytechnic Institute
Drake, Michael, University of California, Irvine
Dressen, Tiff, University of California, Berkeley
Ehmann, Kornel, Northwestern University
Eiler, Kathleen, University of California, Irvine
Eisenbraun, Eric, University at Albany-SUNY
Ellis, Jonathan, University of Rochester
Emerling, David, Ohio State University
Ervin, Sarah, University of California, Los Angeles
Fancher, Michael, College of Nanoscale Science and Engineering, Albany University
Farland, William, Colorado State University
Fedder, Gary, Carnegie Mellon University

Feng, Philip, Case Western Reserve University
Figurelle, Wayne, The Pennsylvania State University
Fonash, Stephen, The Pennsylvania State University
Franklin, Debra, Wichita State University
Frederick, Dick, Rensselaer Polytechnic Institute
Freeman, Harold, North Carolina State University
Gajewski, John, Cuyahoga Community College
Garza, Wanda, South Texas College
Gatewood, David, Irvine Valley College
Giles, Merle, NCSA
Gilroy, Lisa, Binghamton University
Gonzalez, Miguel, University of Texas - Pan American
Green, Tracy, Lorain County Community College Foundation
Guroff, Robert, Brown University
Gurvich, Vadim, National Institute for Pharmaceutical Technology and Education
Hallacher, Paul, The Pennsylvania State University
Han, Si-Ping, California Institute of Technology
Hardt, David, Massachusetts Institute of Technology
Hardwick, Martin, Rensselaer Polytechnic Institute
Hartman, Nathan, Purdue University
Hedin, Kevin, Colorado State University
Helmlinger Ratcliff, Teresa, North Carolina State University
Hicks, Don, The University of Texas at Dallas
Hochbaum, Allon, University of California, Irvine
Hovsepian, Sarah, Massachusetts Institute of Technology
Huang, George, Department of Mechanical and Materials Engineering, Wright State University
Huang, Wenzhen, University of Massachusetts, Dartmouth
Hull, Robert, Rensselaer Polytechnic Institute
Jackson, Dion, USC Center for Economic Development
Johnson, Wayne, California Institute of Technology
Johnson, Curtis, Stony Brook University
Kalevitch, Maria, Robert Morris University
Karlicek, Robert, Rensselaer Polytechnic Institute
Keeton, Leigh, Lorain County Community College Foundation
Kessler, William, Georgia Institute of Technology
Khalili, Pedram, University of California, Los Angeles
Khine, Michelle, University of California, Irvine

Khoshnevis, Berok, University of Southern California
Kido, Horacio, M&AE, University of California, Irvine
Kinsey, Brad, University of New Hampshire
Kintzel, Edward, Western Kentucky University
Kiwiet, Nicoline, Dutchess County Community College
Klimow, Kate, University of California, Irvine
Knotek, Michael, Renewable and Sustainable Energy Institute, University of Colorado Boulder
Koenig, Roger, Michigan State University
Korkolis, Yannis, University of New Hampshire
Kramer, Daniel, The Ohio State University
Krishnamurthy, Ananth, University of Wisconsin – Madison
Kulinsky, Lawrence, University of California, Irvine
Law, Matt, University of California, Irvine
Leo, Donald, Virginia Polytechnic Institute and State University
Leu, Ming, Missouri University of Science and Technology
Leu, Paul, University of Pittsburgh
Levesque, Robert, San Bernardino Community College District
Levin, Jacob, University of California, Irvine
Levinson, Rachel, Arizona State University
Lewandowski, John, Case Western Reserve University
Liang, Steven, Georgia Institute of Technology
Liehr, Michael, College of Nanoscale Science and Engineering
Lin, Gisela, Micro/Nano Fluidics Fundamentals Focus Center
Liu, Chenghsin, California State University, Los Angeles
Lorenzi, Kathleen Reneau, University of Colorado Boulder
Madou, Marc, University of California, Irvine
Maia, Joao, Case Western Reserve University
Maloney, Tom, Connecticut Center for Advanced Technology
Martukanitz, Richard, Applied Research Laboratory at the Pennsylvania State University
Masanet, Eric, Northwestern University
Matijasevic, Goran, University of California, Irvine
Maute, Kurt, University of Colorado Boulder
McCord, Alisa, OCSTEM/Orange County Department of Education
McCready, Mark, University of Notre Dame
McCullough, Roger, University of Nebraska at Omaha
McGinnis, Leon, Georgia Institute of Technology
McGrath, Michael, LASP, University of Colorado

McGuffn-Cawley, James, Case Western Reserve University
McKay, Leslie, Metro State University
McMeekin, Bill, North Seattle Community College
McNulty, Tim, Carnegie Mellon University
Midturi, Swami, University of Arkansas at Little Rock
Mikesell, Sharell, The Ohio State University
Mingareev, Ilya, Townes Laser Institute, University of Central Florida
Minnich, Tom, Robert C. Byrd Institute
Mino, Michael, The Connecticut College of Technology
Mitchell, Stephen, University of Dayton Research Institute (UDRI)
Mitchell-Williams, Dyanne, TSU
Mittal, Manoj, University of Texas at Arlington Research Institute
Moore, Kevin, Colorado School of Mines
Moo-Young, Keith, California State University, Los Angeles
Morse, Jeffrey, National Nanomanufacturing Network
Moskowitz, Michelle, University of California, Berkeley
Muha, Susan, Cuyahoga Community College
Mumm, Daniel, University of California, Irvine
Murad, Sohail, University of Illinois
Murday, James, University of Southern California
Myers, James, Computational Center for
Nanotechnology Innovations at the Rensselaer Polytechnic Institute
Nasr, Nabil, Rochester Institute of Technology
Neale, Virginia, Northwestern University
Netravali, Anil, Cornell University
Newman, George, Front Range Community College
Nowinski, Caralynn, University of Illinois
Omurtag, Yildirim, Robert Morris University
Ozdoganlar, Burak, Carnegie Mellon University
Pasaogullari, Ugur, University of Connecticut
Paul, Brian, Oregon State University
Popovic, Zoya, University of Colorado
Prawel, David, Colorado State University
Radford, Donald, Colorado State University
Ragan, Regina, University of California, Irvine
Rankin, Patricia, University of Colorado, Boulder Campus
Rathbun, Lynn, National Nanotechnology Infrastructure Network
Rennels, Kathay, Colorado State University –
Community & Economic Development

Reyna, Mario, South Texas College
Richardson, Martin, University of Central Florida
Rimnac, Clare, Case School of Engineering
Robertson, Susan, Solon City Schools
Rosen, David, Georgia Institute of Technology
Roth, John, Penn State Erie, The Behrend College
Rupert, Timothy, University of California, Irvine
Samuelsen, Scott, University of California, Irvine
Sanchez, L. Rafael, University of Colorado Denver
Sanfilippo, Matthew, Carnegie Mellon University
Schaaf, Jim, University of California, Davis
Schmidt, Martin, Massachusetts Institute of Technology
Schneider, Dean, Texas Center for Applied Technology
Schorr, Herbert, University of Southern California
Schwam, David, Case Western Reserve University
Sczechowski, Jeffrey, University of Colorado at Boulder
Sellards, Cynthia, Robert C. Byrd Institute
Shariat, Parvin, California State University, Long Beach
Sharma, Swati, University of California, Irvine
Shenai, Krishna, University of Toledo
Shephard, Mark, Rensselaer Polytechnic Institute
Shin, Yung, Purdue University
Simoneau, Robert, Keene State College
Sirinterlikci, Arif, Robert Morris University
Smith, Kevin Scott, The University of North Carolina at Charlotte
Sobczak, Bruce, Irvine Valley College, ATEP
Somu, Sivasubramanian, Northeastern University
Spivey, Rich, Ohio Manufacturing Institute
Springs, Stacy, Massachusetts Institute of Technology
Steele, Scott, University of Rochester
Streit, Dwight, University of California, Los Angeles
Sun, Lizhi, University of California, Irvine
Sunday, Doug, Lehigh University
Tackett, Ed, University of California, Irvine
Tang, William, University of California, Irvine
Taub, Alan, University of Michigan
Taylor, Patricia, Thomas Nelson Community College
Theyel, Gregory, California State University
Thompson, Daryl, Salt Lake Community College

Thomson, J. Michael, Cuyahoga Community College
Timmer, Douglas, University of Texas – Pan American
Tinderholm, Duane, CCEFP – University of Minnesota
Torrani, Robert, The Connecticut College of Technology
Tralli, David, California Institute of Technology
Tran, Julie, Smart Manufacturing Leadership Coalition, University of California, Los Angeles
Troxell, Wade, Colorado State University
Tsai, Stephen, Stanford University
Turner, Cameron, Colorado School of Mines
Valdevit, Lorenzo, University of California, Irvine
Vangsness, Jean, University of Massachusetts Lowell
von Maltzahn, Wolf, Rensselaer Polytechnic Institute
von Oehsen, Barr, Clemson University
Vosler, Lynn, Front Range Community College
Vukich, John, Pueblo Community College
Walczyk, Daniel, Rensselaer Polytechnic Institute
Walker, Loren, University of Massachusetts
Wallace, Darrell, Youngstown State University
Wang, Su Su, University of Houston
Wang, Ben, Georgia Institute of Technology
Washington, Gregory, University of California, Irvine
Watkins, James, University of Massachusetts
Webster, Ben, University of Colorado Boulder
Weimer, Alan, University of Colorado
Wen, John, Rensselaer Polytechnic Institute
Werle, Kathleen, Irvine Valley College
Wicks, Frank, Edison Tech Center
Williams Jr, Jimmy, The University of North Carolina at Charlotte
Williamson, Keith, Virginia State University
Wilson, Janet, University of California, Irvine
Wilson, Sarah, Workforce Boulder County
Wright, Paul, Center for Information Technology Research in the Interest of Society, University of California, Berkeley
Wright, Michael, University of Massachusetts Amherst
Yagoobi, Jamal, Worcester Polytechnic Institute
Yehiely, Fruma, Northwestern University
Yu, Xiong, Case Western Reserve University
Yu, Tzu-Yang, University of Massachusetts Lowell

Ziegert, John, University of North Carolina at Charlotte
Ziehl, Paul, University of South Carolina
Zorman, Christian, Case Western Reserve University

## Industry (264)

BAE Systems Land & Armaments
BioDevice Design
Boundary Systems
DuPont
Eastman Kodak
General Dynamics - Ordnance and Tactical Systems
Liquidia Technologies
Lumetrics
M7-Technologies
Modria Inc.
Mongan Intech Corporation
Optimax
Procter & Gamble
QED Technologies
Robotics Innovation LLC
Rockwell Automation
RP+M
SCRA Applied Research & Development
Stefan Sydor Optics
Abraham, Margaret (Meg), The Aerospace Corporation
Ahbe, Brad, Canton Drop Forge
Albers, Tracy, GrafTech International Holdings Inc.
Albin, Mary, Westinghouse
Alexander, William, Qualtek Manufacturing, Inc.
Alexander, Chuck, Solid Concepts, Inc.
Alexopoulos, Nick, Broadcom Corporation
Alford, Charles, UES, Inc.
Alvarado, Igor, National Instruments Corporation
Alt, Michael, Eastman Kodak Company
Amis, Eric, United Technologies Research Center
Anderson, Brian, Pratt & Whitney Rocketdyne
Antonsson, Erik, Northrop Grumman Aerospace Systems

Apanius, Matt, SMART Commercialization Center for Microsystems
Aronson, Jesse, Worksystems, Inc.
August, Zachary, Automated Dynamics
Ba, Saffate, G2G Consulting
Bibik, Stacey, Stacy Machine & Tooling, Inc.
Biller, Stephan, General Electric Company
Bjurlin, David, LAB Corperation
Black, Sharon, Blacksmith Multimedia, Inc.
Bolcavage, Ann, Rolls-Royce Corporation
Boyce, Scott, The Dow Chemical Company
Bransch, Harald, Yaskawa
Burdick, Bill, GE Global Research
Burney, Brian, Oliver Manufacturing Co., Inc.
Burns, Ralph, Business Growth Services, LLC
Buskens, Rick, Lockheed Martin Advanced Technology Labs
Cameron, Chuck, RadTech International
Carnahan, Dan, Rockwell Automation
Casebolt, Eric, Cell Sign Technologies
Chalamala, Babu, MEMC Electronic Materials, Inc.
Chong, Dianne, The Boeing Company
Christie, Cynthia, Christie Consulting, LLC
Collins, Sunniva, Swagelok Company
Compton, Owen, DuPont
Coniglio, Philip, Long Island for Technology
Cook, Harold, Tooling Innovations, Inc.
Cotchen, Donald, McGraw-Hill Construction
Craig, Steven, UniControl Inc.
Crouch, Drew, Ball Corporation
Daggon, James, Rice Lake Weighing Systems
Decker, Doug, Northrop Grumman Corporation
Deckert, Curt, Curt Deckert Associates Inc.
Dinsmore, Jason, Dinsmore & Associates, Inc.
Ditchendorf, Charles, CIMdata, Inc.
Dolan, Benjamin, RapidTech
Dotson, Dennis, Dotson Iron Company
Drumm, Gregory, Nation Grinding Inc./Nation Coating Systems
Ducoin, David, Lockheed Martin Electronic Systems
Duggan, Jim, Edison Tech Center
Dykstra, William, Temper

Easterbrooks, Don, Canton Drop Forge
Edwards, Doyle, Brewer Science
Elter, John F, Sustainable Systems LLC
El-Wardany, Tahany, United Technologies Research Center
Esman, Ronald, MITRE Corporation
Feiereisen, William, Intel
Fish, Chris, McAllister & Quinn
Flite, William, Lockheed Martin
Frank, Melissa, FirstEnergy
Freeze, Brent, Sorlox Corporation
Fulton, Michael, Surface Optics Corporation
Galiazzo, Michael, Regional Manufacturing Institute
Gardner, Brian, Chomarat
Garone, Joe, Long Island for Technology
Gelsomini, Tony, Cray
Gies, Jason, Firehole Composites
Gonzalez, Ray, TGD
Goswami, Kisholoy, InnoSense LLC
Graham, Lawrence, PCCAirfoils
Green, Albert, Kent Displays, Inc.
Greenspon, Judy, NPI Services, Inc.
Grimes, Thomas, LNE Group LLC
Grischuk, Walt, Ora Technology
Guerra, Nick, Perfekta Aerospace
Gurumurthy, Prasanna, LuK USA
Gustafson, Daniel, NextEngine
Guymon, Lance, Wolf Robotics
Hackmon, Ernest, Topgallant Energy
Hagerty, Brian, Reactor Institute
Hainsey, Robert, Electro Scientific Industries
Hanes, Hugh, Materion Brush Inc.
Harney, Paul, Perkins+Will
Harricharran, Kunwar, Battelle Production and Field Support
Hartney, Christopher, Jacobs Technology
Hatch, Mark, TechShop
Hatkevich, Steve, American Trim, LLC
Havelock Jr., Samuel, Federated Precision, Inc.
Hayashi, Steven, GE Global Research
Hayashigawa, Don, NxGen Electronics

Henesey, Michael, IBM
Hildreth, Tom, Hildreth & Associates
Hines, Roberta, SGL/HITCO CARBON COMPOSITES
Hockenberger, Glenn, Lockheed Martin
Holcomb, Curtis, SCRA
Hornquist, Edwin, SCE
Horton, Sam, Strategic Marketing Innovations (SMI)
Hovsepian, Sarah, ASRC Research & Technology Solutions
Howes, Christopher, Timken Company
James, Steve, Pratt & Whitney Rocketdyne
Johnson, Andrew, Advanced Technology Consulting, Inc.
Kan, Pamela, Bishop-Wisecarver Corporation
Kania, John, Applied Materials, Inc.
Karpuk, Michael, TDA Research, Inc.
Kar-Roy, Arjun, TowerJazz
Kazeminy, Assad, Irvine Pharmaceutical Services
Kelemen, Marc, NanoSynopsis.com
Kelley, Danna, SCRA Applied R&D
Kennedy, Michele, Fox Run Systems & Solutions
Kerzicnik, Ernest, Enginuity, LLC
Khan, Asad, Kent Displays, Inc.
Kingscott, Kathleen, IBM Research
Knowles, Shawn, Ball Corporation
Kolarik, Robert, The Timken Company
Kolwey, Neil, SWEEP
Kovach, Joseph, Parker Hannifin Corporation
Kuhn, Howard, Exone Company, LLC
Kumar, Raj, Viasystems Group, Inc.
Kurtoglu, Tolga, Palo Alto Research Center
LaRiviere, Don, The Boeing Company
Lawrence, Carl, Swift Tram, Inc.
Lebow, Jeff, Southern California Edison
Lemon, Mike, ITI
Levine, Leanna, ALine, Inc.
Lewis, Karen, Forging Foundation
Licata, Sam, Cummins, Inc.
Logan, Trent, The Boeing Company
Lohr, David, Commonwealth Center for Advanced Manufacturing
Lombardo, Dale, GE Energy

MacDonald, Antonia, UES, Inc.
Malhotra, Girish, Epcot International
Markham, Richard, Polymer Ohio, Inc.
Marshall, Blake, SRA International
Matlik, John, Rolls-Royce Corporation
Maxwell, James, Dynetics
Mayewski, Dave, Rockwell Automation
McClellan, Ken, Ingersoll Machine Tools, Inc.
McCune, Robert, Robert C. McCune & Associates, LLC
McDonald, Kenny, Columbus 2020
McElroy, James, iNEMI
McGough, Mark, Ioxus
McGowan, Scott, Solid Concepts, Inc.
McNab, Michael, Lockheed Martin - Aeronautics
Menassa, Roland, General Motors
Mintz, Jori, LNE Group
Misra, Ashutosh, ITN Energy Systems, Inc.
Moncrieff, Laurie, Adaptive Manufacturing Solutions
Morgan, Marianne, BASF Corporation
Mueller, Larry, CTD Innovation, LLC
Munshi, Naseem, CTD
Natzic, Walt, Grace Technologies
Nelson, Mike, NanoInk, Inc.
Nesi, John, Rockwell Automation
Newhouse, Bill, Canton Drop Forge
Nieves, Erik, Yaskawa Motoman Robotics
Norman, Donald, The Nielsen Norman Group
O'Connor, Gregory, Amalgam Industries, Inc.
O'Day, Paul, American Fiber Manufacturers Assn.
Oliva, Joanna, Samsung C&T
Orme, Steven, Tooling Innovations, Inc.
Pak, Sun, iComputer
Palmintera, Diane, Innovation Associates
Papu, Mareta, UAW-LETC
Patibandla, Nag, Applied Materials, Inc.
Patterson, Clark, RP+M
Pelc, Daryl, The Boeing Company
Pentlicki, Joseph, Oliver Manufacturing Co., Inc.
Peretti, Michael, GE Aviation

Philippi, Therese, Alion Science and Technology
Phillippy, Bob, Newport Corporation
Philpotts, Alvin, Emerging Technology Applications Center (ETAC)
Poliks, Mark, Endicott Interconnect Technologies, Inc.
Polizzi, Anthony, Rocket Science Tutors
Powell, Katie, Munro Companies, Inc.
Rahman, Derrick, SunVeil Solar Inc.
Ralston, Dennis, KLA-Tencor Corporation
Ramsey, Dougals, Alcoa Inc.
Razum, Joe, GE
Rehkopf, Jackie, Plasan Carbon Composites
Rhoda, Doug, Wolf Robotics
Rini, John, LNE Group LLC
Rivard, Adam, Pratt & Whitney
Roberts, Anthony, Jacobs Technology
Robinson, Ron, IT Data Storage / 3DCAD Printer Corproation
Robinson, Jim, TechShop, Inc.
Rockstroh, Todd, GE Aviation
Rothrock, Ginger, RTI International
Rowlen, Kathy, InDevr
Rustic, Joseph, CTL Engineering
Salihagic, Selma, Ball Aerospace & Technologies
Sanne, Anil, Avittor International Corporation
Sariri, Kouros, FMI
Saunders, Glenn, Upstate New York LRIG, Inc.
Schilling, Patricia, Tugboat Software Inc.
Schorzman, Derek, Liquidia Technologies, Inc.
Schramko, Ken, Lam Research
Schroeder, William, Kitware, Inc.
Shattuck, Jay, Underground Systems, Inc.
Sheet, Lubab, Lam Research Corp
Shiveley, Tom, Innovative Industries
Shrader, Eric, Palo Alto Research Center
Silvashy, Ryan, Falcon Foundry
Singh, Prabhjot, GE Global Research
Smith, Robert, Analytical Mechanics Associates
Smith, Gary, D'Addario & Company, Rico International Division
Snell, Michael, Silfex Corporation
Speir, Ryan, Allineate

Srinivas, Girish, TDA Research, Inc.
Stangle-Castor, Nannette, Fuentek, LLC
Stark, Paul, Manaex
Stellon, Gene, Covidien
Stephen, Jennings, Raydiance
Stimson, Tom, The Timken Company
Strong, Amy, Fuentek, LLC
Sullivan, Rose Ann, TechVision21
Sutton, Craig, Deere & Co
Tackett, Edward, RapidTech, Nat'l Center for Rapid Technologies
Thompson, Dave, AlloSource
Tredway, William, United Technologies Research Center
Tymkiw, Andy, Edwards Lifesciences
Ulintz, Peter, Anchor Manufacturing Group
VanDyne, Ed, VanDyne SuperTurbo, Inc.
Vargas, John, Futek Advanced Sensor Technology, Inc.
Venkataraman, Bala, Magni-Power Company
Veres, Janos, Palo Alto Research Center
Viens, Daniel, United Technologies Research Center
Voss, Bob, Panduit Corporation
Vu, Truc, Microsemi Corporation
Wahlig, Bill, Long Island for Technology
Walker, Rae, Axiom Medical Inc.
Welch, Ken, MSC Software
Wenner, Chauncey, United Launch Alliance
Weyer, Greg, LuK USA LLC
White, Christian, iComputer
Willnecker, Gerald, Powdermet Inc.
Winslow, Kyle, McAllister & Quinn
Wolf, Christopher, ITN Energy Systems, Inc.
Wong, Brian, Enevate Corporation
Wosczyna-birch, Karen, Center for Next Generation Manufacturing
Wu, Yutong, APIC
Yamakage, Masahiro, Toyota Motor Engineering &
Manufacturing North America, Inc.
Yoss, Mark, Lockheed Martin
Zimmer, Stephen, USCAR
Zimmerman, Burr, Urban Venture Group, Ltd.

## Economic Development Organizations (47)

Agenda 2020 Technology Alliance of the Forest Products Industry
Colorado Workforce Development Council
Manufacturing Advocacy and Growth Network
Mid-Columbia Economic Development District
NorTech
Bagley, Rebecca, NorTech
Baunach, Dorothy, Greater Cleveland Partnership
Berger, Scott, AIChE
Black, Sharon, Blacksmith Multimedia, Inc.
Boulier, Paul, Team Northeast Ohio
Brancato, Jeffrey, NorTech
Britton, Patrick, NOCHE – Northeast Ohio Council on Higher Education
Chase, Joan, Greater Cleveland Partnership
Coats, Jack, Center of Innovation for Nanobiotechnology (COIN)
Esoda, Eric Joseph, Northeastern Pennsylvania
Industrial Resource Center, Inc.
Fowler, Linda, Regionerate LLC
Franchell, Michael, Community Based Business Incubator Centerm, Inc.
Gattozzi, Nick, Greater Cleveland Partnership
Giles, Merle, NCSA
Goodpasture, Tim, City of Wichita
Gundersen, Daniel C, Baltimore County Department
of Economic Development
Hoag, Michael, WIRE-Net
Holsheimer, Alex, County of San Bernardino EDA
Houldin, Joseph, Delaware Valley Industrial Resource Center
Hutchison, Hutch, High Technology Rochester
Johnson, Kyle, Next Generation Economy Center
Kedar, Shilpa, The Cleveland Foundation
Klonsinski, Mike, Wisconsin Economic Development Corporation
Knepell, Angie, Colorado Association of Commerce and Industry
Lawrence, Jeffrey, Center for Economic Growth
Lecz, Alan, Workforce Intelligence Network
Lee, George, Glimmerglass Ltd.
Lund, Ken, Colorado Offce of Economic Development
and International Trade
Martell, Terrence, Akron Global Business Accelerator

Martin, Nancy, Monterey Bay Economic Partnership
Milbergs, Egils, Washington Economic Development Commission
Nafziger, Wendi, Longmont Area Economic Council
Nance, Christopher, Greater Cleveland Partnership
Paytas, Jerry, Fourth Economy
Pfunder, Jack, Manufacturers Resource Center
Pritchard, Peter, Center for Economic Growth
Reinhardt, William, Edison Tech Center
Robey, Claudette, Next Generation Economy Center
Severson, Peggy, Denver Offce of Economic Development
Shakour, Elias, Michigan Economic Development Corporation
Steinmetz, Joe, APS-Development
Stumo, Michael, Coalition for a Prosperous America

## Federal/State Government (141)

Colorado Offce of Economic Development and International Trade
Naval Surface Warfare Center – Corona Division
North Carolina Department of Commerce
Oak Ridge National Laboratory – Measurement Science & Systems Engineering Division
Aitcheson, Leslie, Army NVESD
Adams, Richard, National Renewable Energy Laboratory
Alexander, Ronald, NASA Glenn Research Center
Anderson, John, CMTC
Anselmi, Robert, Department of Veterans Affairs
Balder-Froid, Kristin, Lawrence Berkeley National Laboratory
Barkman, William, Y-12 National Security Complex
Bartolotta, Paul, NASA Glenn Research Center
Barton, Russell, National Science Foundation
Baumann, Eric, NASA Glenn Research Center
Blue, Craig, Oak Ridge National Laboratory
Boeman, Ray, Oak Ridge National Laboratory
Brennan, Ann, National Renewable Energy Laboratory
Brinkley, David, Offce of the Secretary of Defense
Brown, Sherrod, Senate
Bugnitz, Tom, CAMT
Cantwell, Elizabeth, Lawrence Livermore National Laboratory

Carrillo, Larry, Sandia National Laboratories
Christensen, Dana, National Renewable Energy Laboratory
Christensen, Michael, OSD ManTech
Christensen, John, Offce of Secretary of Defense
Christodoulou, Leo, Advanced Manufacturing Offce
Department of Energy
Cooper, Kenneth, NASA Marshall Space Flight Center
Cox, Ashley, NYSERDA
Cupples, Chris, Offce of Congressman Tim Ryan
Dalzell, Jack, NASA Glenn Research Center
Daniels, Edward, Argonne National Laboratory
Daniels, Lynn, Department of Energy
Dempsey, Patrick, Lawrence Livermore National Laboratory
Duggan, Maureen, Congresswoman Louise Slaughter
Duly, Diane, NASA Glenn Research Center
Eberle, Cliff, Oak Ridge National Laboratory
Elledge, Raymond, CMTC
Evans, Jamey, Department of Energy
Evans, Heather, NIST
Fielding, Jennifer, Air Force Research Laboratory
Gallagher, Patrick, NIST
Gannon, Jesse, Senator Sherrod Brown's Offce
Garver, Lori, NASA Headquarters
Giovannetti, Dean, NASA Ames Research Center
Gomez, Fabio, CMTC
Gove, Jeffrey, Ohio Board of Regents
Green, Virginia, CMTC
Greer, Frank, NASA Jet Propulsion Laboratory
Gupte, Prasad, NIST
Hanigan, William, Medina County Workforce Development
Hardin, John, Offce of Science and Technology
Hardy, David, Department of Energy
Harris, Stewart, NASA Langley Research Center
Hart, Jack, Offce of Advocacy/ Small Business Administration
Heinrichs, Patricia, Air Force
Henschel, Gregory, Department of Education
Herling, Darrell, Pacific Northwest National Laboratory
Hines, John, NASA Ames Research Center
Hogan, Kathleen, Department of Energy

Holesinger, Terry, Los Alamos National Laboratory
Horseman, Ken, Agency of Commerce & Community Development
Hoyt, Diana, NASA
Ivester, Robert, Department of Energy
Kadtke, James, National Nanotechnology
Coordinating Offce
Kaplar, Robert, Sandia National Laboratories
Kathe, Eric, U.S. Army
Kelley, Michael, NIST
Kilbane, Grace, Department of Labor Employment
and Training Administration
Koppel, Jacques, CEMC
Korsmeyer, David, NASA Ames Research Center
Kramer, Bruce, National Science Foundation
Kumar, Abhai, DPAC
Lee, Jonathan, CMTC
Liang, Ranty, NASA Jet Propulsion Laboratory
Liby, Alan, Oak Ridge National Laboratory
Lugo, Ray, NASA Glenn Research Center
Maher, Angela, HUD – Strong Cities, Strong Communities
May, Deborah, Lawrence Livermore National Laboratory
McIlwain, Michael, Idaho National Laboratory
Meador, Michael, NASA Glenn Research Center
Miller, Jason, White House National Economic Council
Miller, Dennis, Y-12 National Security Complex
Min, Sangkee, Lawrence Berkeley National Laboratory
Molnar, Mike, NIST
Narotski, Bill, Ohio EPA – OCAPP
Ott, Ron, Oak Ridge National Laboratory
Peck, Mason, NASA
Peterson, Eric, Idaho National Laboratory
Prusha, Stephen, NASA Jet Propulsion Laboratory
Prymak, Bill, Department of Energy
Pugliano, Victor, US Army ARDEC
Pye, Miriam, New York State Energy Research and
Development Authority
Ratcliff, Adele, OSD (MIBP)
Reinfurt, Edward, Empire State Development Corporation
Reyes, Matthew, NASA Ames Research Center

Richards, Joni, NASA
Rinker, Mike, Pacific Northwest National Laboratory
Roberts, Jim, NYSTAR
Robinson, Kem, Lawrence Berkeley National Laboratory
Russell, John, Air Force Research Laboratory
Sastri, Bhima, Department of Energy
Schen, Michael, NIST
Schmid, Steven, NIST
Scotti, Stephen, NASA Langley Research Center
Shaw, Robert, NASA Glenn Research Center
Sherman, Aleta, CAMT
Shipers, Larry, Sandia National Laboratories
Shkel, Andrei, DARPA
Silcox, Brett, NASA Headquarters
Simmons, Jerry, Sandia National Laboratories
Singerman, Phillip, NIST
Slobodian, Cheryl, CMTC
Smith, Marilyn, US House of Representatives
Sofos, Marina, Advanced Manufacturing Offce, Department of Energy
Sorensen-Bain, Sumer, CAMT
Sosa-Mallory, Michelle, National Renewable Energy Laboratory
Steitz, David, NASA
Streiffer, Stephen, Argonne National Laboratory
Supola, Neil, US Army NVESD
Tartz, Karla, State of Colorado
Tate, LaNetra, NASA
Teter, David, Los Alamos National Laboratory
Thompson, Gary, NIST MEP
Tibrea, Steven, Savannah River National Laboratory
Tolbert, Carol, NASA Glenn Research Center
Trease, Brian, NASA Jet Propulsion Laboratory
Turner, Clayton, NASA LARC
Ulsh, Michael, National Renewable Energy Laboratory
Utz, Robin, US Department of Education
Vaia, Richard, Air Force Research Laboratory, Materials and Manufacturing Directorate
Veris, Kim, NASA Glenn Research Center
Vickers, John, NASA
Visconti, Kelly, Advanced Manufacturing Offce, Department of Energy

Wach, Steve, Savannah River National Laboratory
Ward, Brynne, DRC (OSD ManTech)
Watson, Jim, CMTC
Werkheiser, Mary, NASA Ames Research Center
Werkheiser, Niki, NASA
Wolfenbarger, Debora, NASA Jet Propulsion Laboratory
Wunsch, Thomas, Sandia National Laboratories
Zucca, Gregory, Cuyahoga County Department of Development

## Research & Non-Profits (85)

Center of Innovation for Nanobiotechnology (COIN)
Conexus Indiana
Optoelectronics Industry Development Association
SEMATECH
Semiconductor Equipment and Materials International (SEMI)
Smart Manufacturing Leadership Coalition
Society of Manufacturing Engineers
The Maritime Alliance
The Ohio Aerospace Institute
The Security Network
The Solar Energy Consortium
Adams, James, Metal Powder Industries Federation
Aller, Mike, Space Coast Energy Consortium
Anderson, Steve, SPIE
Balachandra, Anita, SEMATECH/TechVision21
Berry, Daniel, MAGNET
Boccanfuso, Anthony, The National Academies
Brown, G. Ronald, Agenda 2020 Technology Alliance
Carnes, Kelly, TechVision21
Ciesinski, Michael, FlexTech Alliance
Colbert, Beth, Ohio MEP
Conrardy, Chris, EWI
Crocker, Judith, MAGNET
Crumrine, Susan, Southwest Research Institute
Cupoli, Edward, SEMATECH
Dale, James, Metal Powder Industries Federation
Davis, Jonathan, SEMI

Dumitriu, Monica, Ohio Aerospace Institute
Evans, Paul, Southwest Research Institute
Flannigan, Clay, Southwest Research Institute
Fleegle, Gary, National Center for Defense Manufacturing and Machining
Fralix, Michael, [TC]2
Fry, Michele, Young Innovators' Society
Gibson, Kimberly, EWI
Gilmore, Randy, National Center for Manufacturing Sciences
Girard, Jamie, SEMI
Goodstein, Mark, Cleantech Los Angeles
Hardy, Roy, Forging Industry Association
Heaton, Tim, Colorado Advanced Manufacturing Alliance
Hock, Carolyn, InSoCal CONNECT
Holladay, Dan, SEMATECH – PVMC
Horton, Nancy, Energy Industries of Ohio
Hunt, Warren, The Minerals, Metals & Materials Society
Hurt, Pam, Society of Manufacturing Engineers
Kelbassa, Ingomar, Fraunhofer ILT
Krizman, Greg, MAGNET
Lanke, Eric, National Fluid Power Association
Looman, Kathleen, Society of Manufacturing Engineers
Majcher, Donald, Ohio Aerospace Institute
Maloney, Thomas, Connecticut Center for Advanced Technology
Martin, Michael, Edison Materials Technology Center (EMTEC)
McAulay, Jeff, Fraunhofer CSE
McDougle, Jack, Council on Competitiveness
McIntyre, Cynthia, Compete. Council on Competitiveness
McKinnis, Leonard, Center for Labor and Community Research
Meyer, Carl, The Solar Energy Consortium
Mudry, Rosemary, Energy Industries of Ohio
Nolan, Edward, MAGNET
O'Donnell, Michael, MAGNET
Otterman, Nadine, Young Innovators' Society
Pacelli, Mary Ann, MAGNET
Porter, Kweli, Rainforest Recycling Company
Rael, Denise, FlexTech Alliance
Rafferty, Bill, Southwest Research Institute
Rankin, Tyra, Texas Clean Energy Council
Resnick, Ralph, National Center for Manufacturing Sciences

Rochester Regional Photonics Cluster
Ruen Blanchard, Sarah, ASERTTI
Salay, David, Ohio Aerospace Institute
Sattler, Dwayne, Ohio Technology Consortium (OH-TECH)
Schmidt, Bob, MAGNET
Schober, John, MAGNET
Sloan, Susan,The National Academies - GUIRR
Smith, Larry, SEMATECH
Snow, Dave, Purdue Manufacturing Extension Partnership Center
Srivastava, Anil, TechSolve, Inc.
Taylor, Rebecca, National Center for Manufacturing Sciences
Thompson, Dennis, SCRA
Tieckelmann, Robert, SEMATECH
Vairamohan, Baskar, Electric Power Research Institute (EPRI)
Vaze, Suhas, EWI
Warndorf, Paul, AMT – The Association For
Manufacturing Technology
Wellington, Jane, Society of Manufacturing Engineers
Weil, Richard, Cleantech Los Angeles
Whitefoot, Katie, National Academy of Engineering
Wong, Perry, Milken Institute

## Other (27)

Chris Cassidy Consulting
JumpStart
The Information Technology & Innovation Foundation
The Motorsports Education Foundation
Berkowitz, Kevin
Buthe, Chris, Lean Enterprise Development
Denning, Mary Kaye, Capital of Know-How
Gartenlaub, Marshall, Technical Employment Training
Gary, Steve, Avery Dennison
Grover, Kristie, BIOCOM Institute
Hanmin, Chen
Hooper, Nicholas, UK Science and Innovation Network
Huggan, Phillip
Jory, Rick

Kakiris, Barbara, Barbara Kakiris Consulting, LLC
Kerzicnik, Ernest, Enginuity, LLC
Loveman, Stanford, Out of the Blox
Marcal, Pedro, MPACT
Mouakkad, Sally, British Consulate-General Los Angeles
Overmoyer, Richard, Fourth Economy Consulting
Payne, David, Ohio Republican Party
Petersen, Kevin, Food Chain Safety
Reyes, Matthew, Exploration Solutions, Inc.
Rojhani, Ethan, PwC
Talnack, Marie, Talnack & Associates
Trerotola, Ron, Trerotola Associates
Tseng, Shirley

## APPENDIX B. GLOSSARY OF ABBREVIATIONS

| | |
|---|---|
| AMNPO | Advanced Manufacturing National Program Office |
| AIR | Accelerating Innovation Research |
| BAA | Broad Agency Announcement |
| CRS | Congressional Research Service |
| DOC | Department of Commerce |
| DOD | Department of Defense |
| DOE | Department of Energy |
| ED | Department of Education |
| ERC | Engineering Research Centers |
| ETA | Employment and Training Administration |
| GDP | Gross Domestic Product |
| GOALI | Grant Opportunities for Academic Liaison with Industry |
| IMI | Institute for Manufacturing Innovation |
| I/UCRC | Industry/University Cooperative Research Centers |
| MDF | Manufacturing Demonstration Facilities |
| MENTOR | Manufacturing Experimentation and Outreach |
| MEP | Manufacturing Extension Partnership |
| MRL | Manufacturing Readiness Level |
| MRSEC | Materials Research Science and Engineering Centers |
| NACK | National Nanotechnology Applications and Career Knowledge |

**Appendix B. (Continued)**

| | |
|---|---|
| NAMII | National Additive Manufacturing Innovation Institute |
| NASA | National Aeronautics and Space Administration |
| NCDMM | National Center for Defense Manufacturing and Machining |
| NIST | National Institute of Standards and Technology |
| NNMI | National Network for Manufacturing Innovation |
| NSF | National Science Foundation |
| PFI | Partnerships for Innovation |
| R&D | Research and Development |
| RFI | Request for Information |
| SBIR | Small Business Innovation Research |
| SME | Small and Medium Enterprises |
| STTR | Small Business Technology Transfer |
| TRL | Technology Readiness Level |

In: National Network for Manufacturing … ISBN: 978-1-63117-681-4
Editor: Vivianna R. Bagnoli 

*Chapter 3*

# TESTIMONY OF PENNY PRITZKER, SECRETARY OF COMMERCE, U.S. COMMERCE DEPARTMENT. HEARING ON "THE ROLE OF MANUFACTURING HUBS IN A 21ST CENTURY INNOVATION ECONOMY"*

## INTRODUCTION

Chairman Rockefeller, Ranking Member Thune, and members of the Committee, thank you for calling this important hearing to examine the role of manufacturing hubs in a 21st Century innovation economy.

I welcome the opportunity to discuss a proposed National Network for Manufacturing Innovation (NNMI), and am supportive of the approach in the bipartisan legislation recently introduced by Senators Sherrod Brown and Roy Blunt on this topic.

The NNMI legislative proposal would largely implement recommendations by the first Advanced Manufacturing Partnership (AMP) Steering Committee, a task force of 12 leading company CEOs and six university presidents, with input from 1,700 members of industry and academia. Co-chaired by Dow CEO Andrew Liveris and former Massachusetts Institute of Technology (MIT) President Susan Hockfield, the AMP Steering Committee issued a report to the President in July 2012 entitled *Capturing*

* This is an edited, reformatted and augmented version of testimony presented November 13, 2013 before the Senate Committee on Commerce, Science and Transportation.

*Domestic Competitive Advantage in Advanced Manufacturing.* Among its findings was the need for a network of manufacturing innovation institutes. These institutes would allow companies to collaboratively invest in precompetitive research to tackle manufacturing challenges they cannot address individually. The institutes would provide companies, including small manufacturers, access to capital equipment and facilities to conduct testing and research in order to accelerate to the market new cutting edge technologies. A new generation of our manufacturing workforce would be trained in an environment similar to a "teaching hospital" for advanced manufacturing, where engineers, researchers, and workers are able to gain new skills and capabilities working on state-of-the-art equipment and new manufacturing challenges.

## MANUFACTURING INNOVATION

It is an exciting vision that has been widely embraced. Just two weeks ago, the National Association of Manufacturers (NAM) and other organizations including Semiconductor Equipment and Materials International, publicly announced that they are supporting this effort.

Before elaborating on this vision, I would first like to discuss the commercial problems these institutes are meant to solve. As you know, I come to the Department of Commerce as a business person, particularly sensitive to market demands. Many in private industry have embraced the institute concept because they see a need for industry leaders to collaborate on advanced technology challenges in order for the United States to secure a competitive edge. Others believe it will spark economic development in regions that have been hard-hit by previous recessions. And these private sector leaders wholeheartedly endorse the Institutes as a critical national investment to rebuild manufacturing capabilities and strengths that have eroded over the last decade as manufacturing went offshore.

In fact, since 2001, the United States has lost production across a range of advanced technology industries where the United States had previously been dominant. This phenomenon was caused largely by companies' increased reliance on global supply chains, which has allowed companies to tap specialized manufacturers from other countries to produce high-performance parts. Over the long-run, this reliance shifted production and often innovation overseas as well.

As a recent MIT study points out, innovation occurs not only at the point of invention, but at every stage of product development and delivery. This was the rationale behind the iconic Bell Labs where engineers co-located beside technicians to develop and continually improve production processes for telecommunications equipment. Or at companies like DuPont, where partnerships between design and production specialists led to an affordable manufacturing process for Kevlar in 1970 — finally making the material marketable, five years after it had been invented.

On the other hand, when a company's inventors and design engineers are separated from the production process, that company may be hindered in its abilities to improve products or develop new goods and services. This separation is why scholars now suggest that certain industries created in the United States — such as flat panel displays and certain consumer electronics — have moved entirely offshore. In some of these cases, manufacturers built assembly facilities in Asia which made it attractive for parts suppliers to re-locate there as well. Soon, entire supply chains were migrating out of the United States, and we lost our ability to lead in the innovation and production of these types of products.

The NNMI will allow the United States to rebuild the dense networks of capabilities that it lost during the past two decades of manufacturing offshoring. Bringing large manufacturers, universities, and small businesses together in institutes will help restore U.S. competitiveness in manufacturing. This is critically important for boosting U.S. innovation and exports, and it will facilitate middle class job growth. Indeed, the manufacturing sector accounts for 70 percent of U.S. private-sector research and development, 70 percent of patents, and the vast majority of U.S. exports. Manufacturing jobs provide a key pathway into the middle class, with workers earning between wages and benefits 17% more than their counterparts in other sectors.

Catalyzing industry to strengthen American manufacturing capabilities has worked for us in the past. Twenty-seven years ago, the Reagan Administration sounded the alarm over a crisis in the semiconductor industry. We were fast losing market share and would eventually lose our entire industry if nothing was done.

For the Department of Defense this posed a national security risk. As a result, SEMATECH was born; the Pentagon invested $500 million into a small consortium of U.S. companies, allowing industry rivals to collaborate in road-mapping the future of semiconductor chip technology and to develop the manufacturing processes necessary to mass produce those chips. Instead of each semiconductor manufacturer spending money to design its own equipment in isolation, SEMATECH provided a forum for companies to work together and develop common standards for next generation chip manufacturing technology.

Just this last July, I visited SEMATECH at its new home on the seven-year old campus of the College of Nanoscale Engineering of the State University of New York (SUNY) in Albany, New York. SEMATECH has not received federal matching funding for over fifteen years, having evolved into a self-sufficient enterprise - and having grown to include over a hundred international players.

What I found particularly fascinating were some of the College's other partners. Unlike SEMATECH, which focuses on research into manufacturing processes, the College's other partners are helping transition additional emerging research into actual manufacturing capabilities — in the same vein that we are proposing for NNMI. One such initiative is the Global 450 Consortium — an effort to make the surface on which we make chips — called wafers bigger. This could bring down costs and add more functions to our smartphones, tablets, and car electronics.

This consortium, started up just in the last two years, is comprised of IBM, Intel, Samsung, TMSC, and Global Foundries as well as the State of New York. Each company recognized this effort was high risk, extremely expensive, and beyond the capability of a single company. But, on this great college campus, these companies are now pooling resources and sharing risk to try to develop manufacturing equipment together. By collaborating they advance the "pre-competitive" technology that can be inserted into final products that are more proprietary. Because this work goes on here, the innovation, company growth, and high tech jobs are here. Additionally, students at various levels on the campus will have an opportunity to train on how to use this equipment and develop new production techniques. The College itself functions as a trusted third party, not only providing the space for conducting collaborative research, but also managing complex arrangements for sharing intellectual property. The College is in effect coach, convener, and arbiter — and has developed a successful model for protecting companies' proprietary interests.

The innovations taking place at this facility are breathtaking. So much so, that semiconductor companies from around the world are investing tens of billions of dollars to build factories in that region and creating thousands of new jobs, just to be close to the College and tap the collaborative research and scholars available nearby. In fact, just this year Global Foundries the second largest chip maker in the world — announced another $7 billion expansion of its multi-billion dollar facility in neighboring Saratoga Springs.

Companies in this industry are flocking to the region because they believe being a part of this industrial ecosystem is critical to their long-term competitiveness; in turn they are building multi-billion dollar complexes. This case shows what can happen when the government whether federal or state - provides initial seed funding, and helps convene industry and university partners to collaborate in manufacturing research and workforce training. In my view, this should be an inspiration for aspiring NNMI institutes.

## THE MISSING MIDDLE

What is transpiring in upstate New York can and must take place elsewhere around our country. The United States has long invested public dollars in initial or basic research, and, in many industries, companies are likely to invest in late-stage development — once a product has been proven and a market is beginning to materialize. *But* what about that middle stage — when a technology has been invented but there is no established process for scaling up its production?

To lead the world in advanced manufacturing means to lead not just in initial invention but all the way through production. Other countries, particularly in Asia and Europe, have been investing billions of dollars in such "technology transition" for decades, and have their own programs analogous to NNMI. It is time to find a uniquely American solution to the challenges associated with moving technology from lab to shop.

## NATIONAL NETWORK FOR MANUFACTURING INNOVATION

A uniquely American solution must be led by the private sector. The Administration relied on the advice of 1,700 members of industry and academia who provided inputs into the AMP report. The NNMI program will eventually be wholly owned and operated by companies and universities — not the Federal Government. However, as evidenced by history, these endeavors will require seed funding to make it possible for companies, universities, community colleges, nonprofits, and others to join a manufacturing innovation consortium. Our proposal is for the government to provide that "patient capital" for about five to seven years, and then allow an institute to operate on its own. In essence, the public investment is in the U.S. innovation ecosystem — to create the space for industry and academia to solve industry-relevant problems. In the process, the institute will need to prove that it can and will be self-sustaining long after the government ends its investment, and that it can meet critical market demands. There is indeed a demand for this sort of program. I have seen it firsthand in Albany. But I can assure you, there is a hunger for these institutes all across the country. This is evidenced by the great interest and robust competitions for new manufacturing innovation institutes being held by the Departments of Defense and Energy right now.

As members of this committee are aware, last year, the Department of Defense led a competition to establish a pilot institute. The institute focuses on additive manufacturing and 3D printing an area of great importance both to the Armed Forces and the broader economy. This technology is literally something out of "Star Trek" — allowing individuals to use a computer to design intricate structures and shapes traditional manufacturing processes simply could not make, and then "printing" or "beaming" them layer-by-layer into existence with unprecedented precision.

The Air Force and Navy want to use this technology to build high-performance aircraft and engine parts. The Army and Defense Logistics Agency might use this technology to have instant access to spare parts in-theater, when combat vehicles break down, or to recreate replacement parts that have been out of stock for decades. And in the commercial market, one can conceive of virtually endless applications for these tools — ranging from automotive and medical device production to fashion and apparel. These sophisticated and expensive production machines are far from perfect — and the institute is a big part of making them better. It is also important to develop the standards for 3D printed parts, training programs, and the skilled workforce needed to support new businesses. Improvements in these production machines are spilling over to consumer markets, where inexpensive printers for schools and homes are rapidly expanding. But much remains to be done to bring the technology fully into the mainstream.

Moreover, if the United States is going to be a global leader in advanced technologies, such as 3D printing, we are going to need to bring all of our leading industry and university resources to bear. We need to recognize that the rest of the world is not sitting on the sidelines. I assure you, many other countries are investing heavily in 3D printing as well as a host of other advanced manufacturing technologies. We need to get this right to remain globally competitive.

Fortunately, there is high demand in the United States to do just that.

From what I am told, competition was fierce for the pilot institute. In the end, the government put $30 million on the table, and the winning consortium matched this sum with an additional $40 million. That money is going towards funding labor, equipment, and applied research projects. The pilot institute was formerly known as the National Additive Manufacturing Innovation Institute, and was recently renamed America Makes. Its headquarters is now open for business in Youngstown, Ohio. Consortium members included, among others, Northrop Grumman, Honeywell, IBM, Timken, and RTI International; university participants included Case Western, Carnegie Mellon, Penn State, and Youngstown State — as well as Marshall University in your home state, Mr. Chairman. And membership continues to grow. While many feel the future promise of this technology is certain to be astonishing, what is far from certain is where the global innovation hub for 3D printing and additive manufacturing will be located. The mission of our pilot institute is to ensure this hub is in the United States.

But what exactly happens at an institute?

The answer is: "A lot." Industry members are crafting detailed roadmaps of their technology needs, collectively defining milestones and then developing strategies to meet those goals. Institute members are collaborating on applied research projects, developing facilities to evaluate nascent technologies and improving equipment and processes for unproven technology to be scaled up to production. By using shared facilities, manufacturers can pool their risk, and drive down the cost of commercialization. Working with university researchers and design engineers, manufacturers can accelerate the insertion of these critical technologies into mainstream manufacturing. Through this process, institute members are establishing new business networks, coordinating their actions, and redefining supply chains and business practices. In particular, institutes offer opportunities for small and medium-sized enterprises to enter these supply chains and access equipment they ordinarily would not be able to afford to use. Finally, community colleges and universities are training new generations of workers on cutting edge technology available at the institute and establishing pipelines for U.S. employers to hire skilled workers. By engaging with colleges and universities, the institutes will both incent and support educators to assure the knowledge and skills of our workforce — building additional science, technology, engineering and mathematic (STEM) career pathways for youth and adults. This aspect of the institutes will be essential to keep the manufacture of new products from moving offshore.

The institute itself provides facilities, equipment, and software for collaborative research. It also helps arrange intellectual property sharing, both to advance technology breakthroughs as well as protect members' proprietary rights. As illustrated below, institute members can range from research universities and national laboratories to community colleges; large manufacturing companies to small and medium-sized enterprises and start-ups; and state and local governments to economic development organizations.

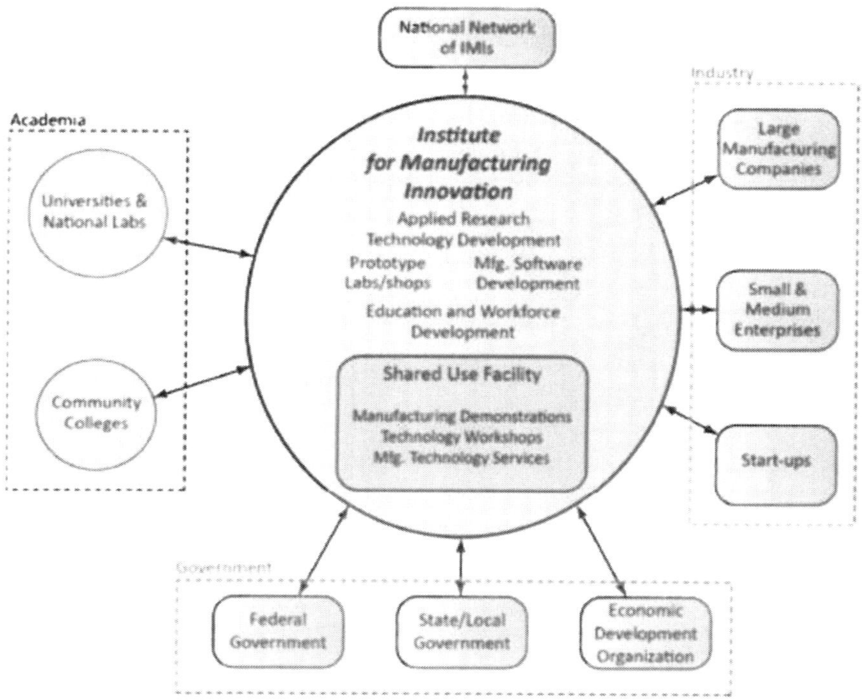


Ultimately, the Administration envisions a collection of these institutes forming a network.

Institutes will thrive not only from collective action within their own consortia, but also through cross-pollination across industries. All parties involved in these institutes will advance competitiveness in their respective technology fields as well as more broadly support our economic and national security interests.

The pilot institute currently underway has demonstrated the great demand for these capabilities among all of these actors. Applied research projects co-financed by the government, industry, and universities are well-underway. And the institute continues to see its membership and private investment rise. For example, in May of this year, Siemens announced a $440 million in-kind grant to Youngstown State University to help train the next generation of 3D printing manufacturers. I am eager to see how it matures over the next several years.

Meanwhile, the Departments of Defense and Energy have launched three more competitions. The Pentagon solicited proposals for consortia focused on lightweight and modern metals as well as digital design in manufacturing. Energy's new institute will focus on enabling more powerful, efficient, and more cost effective power electronics. These are critical endeavors, and we are eagerly awaiting the announcement of competition winners. These new institutes, like the pilot, will continue to help us hone this model and, more importantly, address critical manufacturing technology needs for our country.

In sum, by developing these institutes in partnership with industry and academia, the government will address two critical issues industry leaders say they cannot solve without assistance.

First, it will incentivize companies and universities to work together in promoting innovation and production here at home. Second, it will help companies within an industry ensure that U.S. supply chains are effective and well-integrated.

## CONGRESSIONAL AUTHORIZATION

In order to enable this program to reach its full potential, Congressional action is required to authorize this program and provide the necessary funding. The "Revitalize American Manufacturing and Innovation Act of 2013," sponsored by Senators Blunt and Brown, would authorize Commerce to award funds to assist in planning, establishing, or supporting the centers for manufacturing innovation, which will constitute the network. It would also establish a National Office of the Network for Manufacturing Innovation Program within the National Institute of Standards and Technology to oversee and carry out the program. These are both important steps in meeting the President's call for the NNMI.

Although the institutes being developed under the leadership of the Departments of Defense and Energy under existing authorities are important, at present we have no federal program exclusively focused on identifying emerging technologies with broad potential impact and bringing together companies in associated industries to improve technology transition. Instead, we are meeting our nation's demand by tapping existing federal programs. By relying on these mission agencies, we are confining the institutes' topics to areas that are relevant primarily to these agencies' missions, which exclude many other topics with significant potential impact on manufacturing.

Ultimately, these institutes need to be driven from the bottom up; we need to empower U.S. industries to identify where their comparative advantage lies and where the need is greatest to collaborate in manufacturing innovation institutes.

Our role should be to support those decisions, help them underwrite risk when we can, and thus help unleash the full might of the American innovation economy.

This question was posed to industry in five different public events along with a request for public comment — "What are the topics that are most important to industry, appropriate for an institute, and that industry would co-invest in?" The response was overwhelming — 135 topics. Which should be established? Let consortia teams put together proposals based on market needs — not a federal agency's requirements. Awards will be made through merit-based competition that selects the best business case to receive start-up funding and become part of the network. Additionally, we will continue to rely on the private sector, through forums such as the Advanced Manufacturing Partnership, to ensure NNMI meets industry needs and advances American competitiveness.

In fiscal year 2014, the President's Budget has called for $1 billion for up to 15 institutes built over multiple years to jumpstart the vision shared by AMP, NAM, and so many other leaders across industry. This proposed multi-year funding will provide the consortia the certainty needed for members to commit to fully matching the government contribution. As the concept has developed, so has the demand. In fact, the Administration believes the network could ultimately reach a total of 45 institutes. As a comparison, Germany has 60 *Fraunhofer* Institutes — similar in concept to NNMI — despite having an economy a fraction of the size of ours. It is critically important that we make the NNMI a reality. The approach in the Brown-Blunt bill is an important step in advancing this conversation.

I look forward to working with this committee to move this important legislation forward.

## CONCLUSION

Today, manufacturing remains critical for both economic and national security. As I discussed at the outset, addressing the challenges facing America's manufacturing sector is absolutely essential both for our ability to employ skilled workers as well as for our ability to maintain our competitive edge in the world economy.

Innovation is America's comparative advantage. It is the lifeblood of advanced industries that are fundamental to our nation's future — from nanomanufacturing to cyber-technology. But to truly stay out front in these exciting new fields, it is not enough to invent new products — we can and we must strengthen the ability to make these products too. Establishing the National Network for Manufacturing Innovation is essential to doing just that. I look forward to working with you to realize this vision.

This is just one piece of the Administration's manufacturing agenda; I look forward to returning to talk with you about other exciting initiatives in the near future.

Thank you for the opportunity to appear before you today. I look forward to answering your questions.

In: National Network for Manufacturing … ISBN: 978-1-63117-681-4
Editor: Vivianna R. Bagnoli 

*Chapter 4*

# NATIONAL NETWORK FOR MANUFACTURING INNOVATION FREQUENTLY ASKED QUESTIONS*

***Advanced Manufacturing National Program Office***

1) Announcement of Three New Institutes
2) National Network for Manufacturing Innovation
3) NNMI Pilot Institute
4) The NNMI Institutes
5) Evaluation and Metrics

## 1. MAY 2013 ANNOUNCEMENT OF THREE NEW INSTITUTES

### Q: When Are Agencies Announcing These New Institutes?

- The Department of Defense (DOD) and the Department of Energy (DOE )will be taking the lead to launch three new manufacturing innovation Institutes in 2013 through an open and competitive process. DoD has the lead to establish two Institutes and DOE has the

---

* This is an edited, reformatted and augmented version of a document released by the Advanced Manufacturing National Program Office, dated August 28, 2013.

lead to establish the other Institute. We will release further information about the timeline in the coming solicitations, and expect to award each Institute by the end of 2013.

## Q: How Much Federal Investment Will Go into Each Institute?

- The Obama Administration has announced a Federal commitment of $200 million for the competition on the three new manufacturing Institutes, which will be led by the Department of Defense and Department of Energy in partnership with the Department of Commerce, NASA, and the National Science Foundation. DOD will lead two Institutes, and DOE will lead the other.
- The agencies will provide additional information on the total investment for each new Institute with the solicitations. The NNMI Preliminary Concept Design Paper estimates each Institute's Federal funding contribution will approach $70 million - $120 million. Federal funding for the two DOD Institutes and the DOE Institute will fall within that range.

## Q: What Are the Technology Focus Areas of These New Institutes?

- The two DOD Institutes will be focused on (1) Digital Manufacturing and Design Innovation (DMDI); and (2) Lightweight and Modern Metals Manufacturing Innovation (LM3I). The DOE Clean Energy Manufacturing Innovation Institute will be focused on wide bandgap semiconductor power electronic devices.
- Given that the three new Institutes will be led by DOD and DOE and funded within existing authorities, these Institutes will focus on cross-cutting manufacturing technologies that address critical national security and energy needs.
- The DOD's DMDI Institute will address the life cycle of digital data interchanged among myriad design, engineering, manufacturing and maintenance systems, and flowing across a networked supply chain. Its LM3I Institute will accelerate the introduction and expand the use of more affordable products made with high strength-to-weight alloys that improve performance and reduce energy consumption.

- The DOE's Clean Energy Manufacturing Innovation Institute will focus on the use and manufacture of materials capable of handling high voltage and power loads in a new generation of highly efficient and reliable semiconductor devices that can improve performance of the electric grid, renewable power, electric motors, power server supplies and consumer electronics.

## Q: Will the National Additive Manufacturing Innovation Institute (NAMII) Pilot Be Funded to the Level Currently Envisioned for NNMI Institutes?

- The NNMI Preliminary Concept Design Paper ("National Network for Manufacturing Innovation: A Preliminary Design"), which the Administration released in January 2013, estimated that each Institute's Federal funding contribution would approach \$70 million - \$120 million. Federal funding for the two DOD Institutes and the DOE Institute falls within that range, with \$200 million spread across three Institutes.
- By comparison, the NAMII pilot Institute received an initial Federal investment of \$30 million and a non-Federal funding commitment of \$45 million. NAMII was established as a pilot Institute to help demonstrate the need for the development of the broader NNMI. The NAMII cooperative agreement, however, does provide enough flexibility and contractual ceiling to enable scale-up to the size of a full NNMI Institute.

## Q: Can an Offeror Partner with a Federal Entity, Such As an Agency, National Lab or FFRDC, on a DOD Institute?

- No. The DOD acquisition host is already working with an interagency DOD and civilian agency team and as a signatory to the eventual cooperative agreement is already a partner. For this reason – and the fact that any potential federal partner would have to make itself available to any other proposer, which is impractical -- federal government participating agencies shall not be included on any team responding to a DOD solicitation for the DMDI or LM3I institute.

- This does not preclude solicitation offerors from citing their experience working with other federal activities as a means of demonstrating their ability to work with the DOD and/or civilian agencies. Such references will actually be encouraged.
- Post award, it is plausible that federal agencies, centers or laboratories will be allowed to become members of a DOD Institute if they choose to interact to even a higher degree with that particular institute. That will be a decision left up to the Institute leadership.
- Non-federal entities, including entities who may receive federal funding support, are eligible to participate in an offeror proposal. Such entities include public universities and community colleges, not-for-profits who may have current or prior federal contracts, and Manufacturing Extension Partnership (MEP) centers.

## Q: Can a University Lead a DOD Institute?

- While the recipient of an Institute award can be a research-oriented entity such as a U.S. university or other not-for- profit research center, that university must form and propose an independent non-profit partnership that includes a cluster of manufacturing firms and associated educational and economic development organizations.
- DOD Institutes must be led by independent, not-for-profit institutions that strongly leverage industry consortia, regional clusters, and other resources in science, technology, and economic development. Institutes are intended to link and leverage all available resources, including institutions funded through existing Federal programs, so that they have national and global impact.

## Q: Can Manufacturing Extension Partnerships (MEPs) Participate in the DOD Institute Solicitations?

- The organizations that operate MEPs are eligible to participate in the two DOD Institute solicitations; Digital Design and Manufacturing Innovation (DMDI) and Lightweight and Modern Metals Innovation (LM3I). MEPs are largely independent, but do receive up to 1/3 of their funding from federal sources, specifically from the Department of Commerce, National Institutes of Standards and Technology

(NIST). While MEPs do have a federal affiliation, they are independent with respect to their local engagement. Individual MEP Centers cannot reasonably be expected to be a national resource available to all potential offerors, regardless of locale.

- MEPs are typically run by non-profit organizations. These organizations – not the MEP Center itself -- are eligible to participate in the DOD Institute solicitations (DMDI and LM3I). Such organizations can reference their involvement in managing an MEP Center as part of a proposal.
- That portion of any resources offered in a DMDI or LM3I proposal that are paid for with federal funds contributed to the MEP Center cannot count toward cost share.

# 2. The National Network for Manufacturing Innovation (NNMI)

## Q: What Is the NNMI?

- In January 2013, the Administration released a design document outlining a vision for implementing the President's proposal. http://www.manufacturing prelim design.pdf
- Proposed by President Obama, the National Network for Manufacturing Innovation (NNMI) is a linked network of Institutes to support innovation in manufacturing across the United States.
- The NNMI fills a gap – the "missing middle" - between the nation's R&D, education and training activities and the deployment of technological innovations and a skilled workforce in domestic production of goods.
- At their core, these individual Institutes for Manufacturing Innovation (IMIs) will be world-leading centers that bring together companies of all sizes, universities and other educational institutions, state and local governments, the Federal government, and others to co-invest in the development of technologies that businesses can use to innovate production processes and new products.

## Q: Are There Other Programs Similar to the NNMI?

- The President's vision is influenced by models that have worked in other countries. The importance of manufacturing to innovation and a nation's economy is widely recognized and a number of nations have developed programs that nurture this relationship. For example, Germany, Korea, and Japan each have established government-industry-academia partnerships to spur industrial innovation.
- Descriptions of many of these foreign programs are provided in a 2011 report from the non-partisan Information Technology and Innovation Foundation. Some of these programs include the United Kingdom's Manufacturing Advisory Service; Germany's Fraunhofer Institutes; Taiwan's Industrial Technology Research Institute; Japan's Public Industrial Technology Research Institutes; and Australia's Enterprise Connect.
- Understanding and awareness of these other programs informs the U.S. approach to developing a manufacturing infrastructure, but there is no one-size- fits-all approach that should be applied to the United States without careful consideration of the unique needs and resources of our nation.

## Q: How Does the NIST Advanced Manufacturing Technology Consortium (AMTech) Program Differ from the NNMI?

- The new NIST AMTech Program complements the NNMI. AMTech provides grants to industry-driven consortia to identify and prioritize research projects addressing long-term U.S. industrial research needs. Technology roadmaps and related AMTech-enabled outputs will help guide pre-competitive, infrastructural research, some of which will be supported by AMTech in future years through funding for university and government laboratory research. The NNMI is a network of linked collaborative centers (IMIs) that will support innovation in advanced manufacturing nationwide. The focus of the co-funded IMIs will be on the economically damaging gap that separates invention and discovery from manufacturing scale-up and commercialization. Equally important, IMIs will build workforce skills at all levels and enhance manufacturing capabilities in companies large and small.

Finally, AMTech-enabled outputs may serve as the equivalents of action items or works-in-progress for IMIs.

## Q: How Is the NNMI Different from State or Local Regional Clusters?

- The NNMI is not competing with state or regional clusters. Rather, each Institute within the NNMI would serve as a regional asset that strengthens a state or regional cluster while linking that locale into a nationwide infrastructure to support advanced manufacturing.
- The NNMI will build on the existing capacity of and investment in regional and local clusters across the United States, as well as incentivize the creation of new regional capacity. Each IMI will leverage industry, university, community college, and Federal agency, as well as state and local resources to accelerate innovation and scale up of needed manufacturing capacity, including an agile and skilled workforce. With the additional Federal resources and significant buy-in from industry, an IMI will be able to accelerate investment in industrially-relevant cross-cutting product and process technologies.
- The linkage of IMIs into a national network serves to support and expand their individual impact. However, success will also depend on critical partnerships between IMIs and regional and state groups as well as relevant R&D efforts such as NSF Engineering Research Centers, DOE Energy Innovation Hubs, Federal laboratories, and other existing national assets.

## Q: Don't We Already Have Many Federal Programs That Achieve the Goals of the NNMI?

- The simple answer is no. There are no current Federal programs that have the required attributes to significantly influence the nation's competitiveness and successfully bridge "the missing middle" in the manner and magnitude proposed for the NNMI. A whole-of-government approach is needed.
- The Federal government has a number of highly effective programs in the area of manufacturing research and development (e.g., NSF Industry/University Cooperative Research Centers and Engineering

Research Centers, DOE Energy Innovation Hubs, and other manufacturing demonstration facilities or centers of excellence like those sponsored by the DOD ManTech Program), but these programs tend to be sector- or agency mission- specific, concentrate on early stage R&D, risk and cost reduction for mission critical acquisitions (e.g., DOD), or promote later-stage commercialization.

- These targeted efforts do not possess the size and scope to fundamentally transform the Nation's manufacturing infrastructure.
- The NNMI provides a national infrastructure to effectively integrate research advances, education and training activities, and the deployment of technological innovations that will result in increased domestic production of goods.

## Q: How Did the Federal Government Come up with the Preliminary Design for the NNMI?

- In January 2013, the Administration released a design document outlining a vision for implementing the President's proposal. http://www.manufacturing.gov/docs/nnmiprelimdesign.pdf
- The Advanced Manufacturing National Program Office (AMNPO) conducted an extensive public engagement initiative to communicate with and involve manufacturing broad set of stakeholders in the design of the NNMI program. This included a Request for Information (RFI) in the Federal Register open for a six-month period. During the RFI open period, the AMNPO created an online "wiki" for public comment, and organized and conducted four "Designing for Impact" workshops across the nation.
- The workshops drew participants from across the United States, representing industry, academia, non-profit organizations, state and local economic development organizations, professional associations, and private individuals. Altogether, the workshops and RFI gathered the input from about 900 leaders. Reports were prepared for each workshop as well as on the RFI responses, and the results were made publicly available on the new manufacturing portal.
- Additionally, for communications and outreach, AMNPO leaders were extensively engaged with industry and academic public events, serving as requested as speakers and panelists. In 2012 the AMNPO provided the keynote speaker in 19 major manufacturing conferences

or expositions, communicating the Administration's manufacturing agenda or explaining the NNMI initiative.

- This extensive input was used to develop the "National Network for Manufacturing Innovation: A Preliminary Design," a Federal report issued by the National Science and Technology Council. The report was publicly released at a design review workshop in Huntsville, Ala., on January 16, 2013.

### Q: Does the NNMI Have a Specific Focus or Can Different Institutes under the NNMI Have Widely Varying Scopes and Focus?

- Each Institute will have a unique and well-defined focus area, such as an advanced material, a manufacturing process, an enabling technology, or an industry sector.
- The focus of each Institute will be on integrating capabilities through collaborations at facilities designed and equipped to address cross-cutting manufacturing challenges, yielding solutions that have the potential to retain, expand, or innovate new industrial production in the United States.
- The proposing teams will be driven by the needs of industry and the opportunities created by new technologies.

## 3. The NNMI Pilot Institute

### Q: What Is NAMII? How Does It Relate to the NNMI Program?

- NAMII, the National Additive Manufacturing Innovation Institute, is the Pilot Institute announced in August, 2012, and headquartered in Youngstown, Ohio. NAMII was launched as a demonstration of the NNMI concept under existing agency budgets and authorities. NAMII shares many features with the envisioned IMIs. The technology topic is on the field of Additive Manufacturing.
- Additive Manufacturing is an emergent area, sometimes called Digital Manufacturing or 3D-printing, where parts are directly manufactured

by computer-controlled equipment using data from descriptive geometry files.

- The NAMII proposal was submitted by a consortium led by a non-profit institution (the National Association for Defense Manufacturing and Machining). Although NAMII is still very young, the Institute continues to rapidly grow, and currently has more than 70 partner entities, including large and small companies, non-profit and professional associations, research universities and community colleges participating in the effort
- NAMII's initial Federal investment is smaller in scale than envisioned IMIs. It was formed with a Federal investment of $30 million, which attracted nearly $40 million in co-investment (cash and in-kind) by industry, academia, state and local entities and other organizations. This was a strong indicator of industry and local participation expected in IMIs.
- Additive manufacturing was pre-selected as a topic to reflect agency needs, while NNMI is designed to maximize impact on the manufacturing sector. Topics will not be pre-selected in NNMI; rather proposals will be openly solicited and will be expected to describe how the proposal topic is driven by industry needs and technological opportunities.

# 4. The NNMI Institutes

## Q: What Is an Institute for Manufacturing Innovation?

- The National Science and Technology Council (NSTC) recently published the proposed framework design of these Institutes and the overall network, which describes the mission and objectives. This was arrived at by extensive input from industry and academia by public workshops and open comment. The report is available at: http://manufacturing.gov/docs/NNMI prelim design.pdf
- The Institutes will have common goals but unique concentrations. The many partners in an Institute – including Federal, state and local governments, industry, and academia – will leverage their resources to collaborate and achieve advances in manufacturing innovation.

- The Institutes will focus on the industry-relevant challenges in scaling up new technologies and manufacturing processes. Institutes must be industry-led and have substantial co-investment by industry.
- Altogether, these Institutes will form a network – the NNMI – that creates a new and effective manufacturing research infrastructure for U.S. industry, and creates a framework across the country that nurtures the required American manufacturing workforce.

## Q: Is an Institute Focus Defined by the Government?

- The NNMI design report, "National Network for Manufacturing Innovation: A Preliminary Design" outlines a program concept that the Institute focus areas should be defined by the proposing teams. This concept, strongly supported by public input, is to allow open solicitations rather than government-selected topics.
- The Institutes that are being launched this year, supported by existing programmatic resources, will be on agency-specified topics given mission needs and consistent with the resources being used.

## Q: How Will the Institutes Engage Small Manufacturers?

- Small and medium-sized enterprises (SMEs) are a vital part of the manufacturing sector. SMEs are early adopters of transformational technology, and they are well-positioned to innovate and produce significant job growth.
- The shared facility infrastructure provided by the NNMI will be an important asset for SMEs.
- IMIs will encourage SME participation through
  - Direct outreach to partners and intermediaries that work closely with SMEs
  - Information and services tailored specifically to SMEs
  - Tiered fee structures and allowance of in-kind contributions to facilitate access to small companies
  - Staged licensing of intellectual property, and similar arrangements.

## Q: How Will the Federal Government Work with State and Local Governments in Establishing and Running the Institutes?

- Most Institutes will be led by an independent, U.S. not-for-profit institution with the capacity to lead an industry-wide technology, workforce development, and infrastructure agenda. For not-for-profit led Institutes the preferred model is a consortium with extensive industry participation and leadership.
- Partners in the Institute should include a range of national, state, and local stakeholders. Each IMI proposal will be expected to outline the methods by which decisions will be made, including those decisions related to operations, membership, intellectual property, project selection, funding allocation, and progress toward sustainability. Draft Institute membership and governance agreements should be included in proposals.
- Each Institute should have substantial autonomy from its partner organizations and institutions and should have an independent fiduciary Board of Directors predominantly composed of industry representatives.

## Q: What Is the Ideal Mix of Partners to Compete for a NNMI-Funded Institute?

- The ideal mix of partners will depend on the proposed IMI focus, but should include all relevant stakeholders needed to develop the infrastructure envisioned by the NNMI program.
- Most Institutes will be led by an independent non-profit organization. Significant participation by large industry (anticipated to be a significant co-investor in the IMI), local universities and community colleges, research organizations, career and technical institutions, and other state, regional, and local public and private entities are likely to be valuable members of a successful IMI.

## Q: What Types of Activities Would an IMI Support?

- IMI activities may include, but are not limited to:
  - applied research and demonstration projects that reduce the cost and risk of commercializing new technologies or that solve generic industrial problems;
  - education and training at all levels, from K-12 through professional training;
  - development of innovative methodologies and practices for enhancing the capabilities of and integrating supply-chains; and
  - engagement with small and medium-sized manufacturing enterprises (SMEs).

## Q: How Will the Institutes under the NNMI Be Funded?

- The 2013 report "National Network for Manufacturing Innovation: A Preliminary Design" outlines a notional plan for disbursing Federal funds to an Institute.
- It is anticipated that for most Institutes, the level of Federal funding will be larger at the start, and will become progressively lower as the Institute progresses towards self-sufficiency.
- The non-Federal funding is anticipated to be large at the time of award based on in-kind items such as equipment and buildings. Over time, the funding will shift to project, member fees, user fees, licensing and other sources.
- The Institutes will be expected to have a plan to become self-sustaining after their Federal NNMI funding ends.

## Q: What Constitutes Co-Investment from the IMI Partners?

- The non-Federal co-investment must be tangible, meaningful, and substantial enough in the aggregate to signal strong and committed industry, regional, and local partnership.
- The co-investment requirement is expected to be met with both cash and in- kind contributions. In-kind and cash contributions may arise from any source, but are only counted as co-investment if they come

from non-Federal sources and directly support the function of the Institute.

- The funds may come from the Institute (for example, using revenue from the licensing of intellectual property); the members of the Institute; state, regional, and local sources (such as economic development agencies); private donations; or other non-Federal sources.

# 5. Evaluation and Metrics

## Q: How Will Proposals for Institutes under the NNMI Be Evaluated? Who Evaluates the Proposals?

- The IMI selection process will be managed by the interagency Advanced Manufacturing National Program Office (AMNPO). Participating Federal agencies currently include DOD, DOE, the Department of Commerce and its National Institute of Standards and Technology, the Department of Education, NASA, and the National Science Foundation. In the President's proposal for a National Network for Manufacturing Innovation, the interagency AMNPO team will be responsible for managing an open, competitive selection process and for executing the award process. However, competitions for Institutes that are launched through executive action will be managed by the lead agency providing the initial seed capital for the Institute.
- Proposals received in response to the solicitation(s) will be evaluated competitively by a review team. The review team may include members of the AMNPO, agency partners, and other experts.
- The merit-based selection process may include pre-proposals, site visits, and economic and business plan analyses.

## Q: What Criteria Will Be Used to Evaluate Proposals for Institutes under the NNMI?

- Solicitations for Institute proposals may be staged, and the design and number of solicitations will depend on the availability and timing of funds.
- Detailed selection criteria will be contained in the IMI solicitation(s). It is expected that criteria may include factors such as:
    - The focus on a critical national need or opportunity for U.S. manufacturing
    - The ability of a region to develop partnerships across companies, universities and local government and economic development organizations and to mobilize that region's resources into a coherent plan for industry impact.
    - Proposed activities targeted at the appropriate readiness level (e.g., TRL/MRL 4- 7)
    - The proposed Institute plan to achieve significant impact in technology development and scale up to commercialization, anticipating widespread adoption and links to job creation and broad economic impacts
    - The proposed Institute resources (e.g., personnel, facilities, and participating entities) supporting the plan
    - The level of co-investment from non-Federal entities, and the strength of the plan for sustainability beyond the initial Federal funding
    - The adequacy of the financial plan
    - The level of engagement with SMEs
    - The suitability of shared facilities
    - The level of involvement and quality of the proposed efforts in education and workforce development
    - The adequacy of the governance and oversight model
- It is expected that cooperative agreements will be the primary funding mechanism for NNMI Institute awards, although other types of grants and contracts may also be used. Continuing Federal support will be contingent on co-investment by businesses and other non-Federal entities, and on progress toward sustainable operations.

## Q: How Will the Federal Government Support the Network of Institutes?

- The NNMI will organize a Network Leadership Council composed of representatives of the Institutes, Federal agencies, and other appropriate entities. The Network Leadership Council will actively look for opportunities to leverage existing resources between Institutes.
- Efficient operation of the Network will be facilitated through common policies.
  Common policies facilitate interaction with SMEs, promote collaboration and movement within the Network, and allow IMIs to share services such human resource management. While recognizing the differing needs of various manufacturing sectors, clusters, and ecosystems, the Network will strive, as far as is practical, to maintain common policies with regard to intellectual property, contract research, operations, accountability, and marketing and branding.
- While each Institute may have its own web presence, they will also participate in and link to the AMNPO hosted "Manufacturing Portal" (www.manufacturing). As the network and funded Institutes emerge, this portal will serve to direct interested parties to the resources and capabilities embodied within the overall Network. Content will include information about the focus of each Institute, structure, governance, contacts, annual reports, news, success stories, member information, and more.

## Q: How Will the Success of an Institute under the NNMI Be Determined?

- In the near term the success of an IMI and the continuation of Federal funding will be based on a number of measures including: amount of co-investment attracted, membership in the Institute, project portfolio, success stories, and/or other benchmarks.
- In the longer term, success will be measured by a number of outcome metrics such as measures of workforce health and size, the availability of new advanced manufacturing technologies resulting in new products and processes, increased trade exports, on-shoring of foreign companies to the U.S. and many other economic indicators.

## Q: How Can I Be Involved if I Have a Great Idea Relating to NNMI?

- The Advanced Manufacturing National Program Office (AMNPO) is housed at NIST and serves as the central point of contact for the NNMI efforts.
- More broadly, you will find many resources at www.manufacturing, including links to Federal programs in advanced manufacturing, information about the NNMI, and other publications and resources.
- Any questions or comments about NNMI can be emailed to amnpo@nist.gov. The AMNPO also provides an e-mail alert service that you can sign up for at http://www.manufacturing.gov /contact.html.

In: National Network for Manufacturing … ISBN: 978-1-63117-681-4
Editor: Vivianna R. Bagnoli 

*Chapter 5*

# DRAFT GUIDANCE ON INTELLECTUAL PROPERTY RIGHTS FOR THE NATIONAL NETWORK FOR MANUFACTURING INNOVATION*

*Michael F. Molnar*

## FOREWORD

The interagency Advanced Manufacturing National Program Office (AMNPO) is hosted by the National Institute of Standards and Technology (NIST). Creation of the AMNPO flows from the recommendation of the President's Council of Advisors on Science and Technology (PCAST), in its June, 2011, Report to the President on Ensuring American Leadership in Advanced Manufacturing,[1] that the Federal government launch a concerted, whole-of-government Advanced Manufacturing Initiative. To that end, this interagency office is charged with:

- Convening and enabling industry-led, private-public partnerships focused on manufacturing innovation and engaging U.S. universities, and

* This is an edited, reformatted and augmented version of a document released by the Advanced Manufacturing National Program Office, dated November 2013.

- Designing and implementing an integrated whole-of-government advanced manufacturing initiative to facilitate collaboration and information sharing across federal agencies.

By coordinating federal resources and programs, the AMNPO seeks to enhance technology transfer in U.S. manufacturing industries and help companies overcome technical obstacles to scaling up production of new technologies.

The National Network for Manufacturing Innovation (NNMI) program proposed by President Obama has the goal of advancing American domestic manufacturing.[2]

The program will seek to accomplish this by creating a robust national innovation ecosystem anchored by a network of Institutes for Manufacturing Innovation (Institutes). The NNMI will fill a gap in the innovation infrastructure, allowing new manufacturing processes and technologies to progress more smoothly from basic research to implementation in manufacturing. The NNMI program will have a scale and focus that is unique, and it is built upon concepts of a strong public-private partnership.'

## ABSTRACT

Using a strategy of broad public engagement, in April 2012, the Advanced Manufacturing National Program Office (AMNPO) began collecting input on the National Network for Manufacturing Innovation (NNMI) program design. The collection of information from the public was initiated by a NIST Request for Information (RFI), published in the *Federal Register,*[3] followed by a series of regional workshops sponsored by AMNPO partner agencies and focused on the issues presented in the RFI.

Reports summarizing the responses to the RFI and the comments received at each workshop were published.[4] In January 2013, the *National Network for Manufacturing Innovation: A Preliminary Design* report was published, built upon public input received.[5] This AMNPO document utilizes the information gathered and puts forth draft guidance associated with Intellectual Property (IP) as it relates to the proposed NNMI program, especially as it relates to the sustainability and industry impact of the individual Institutes that will comprise the NNMI.

# REPORT

## A. Background

The Advanced Manufacturing National Program Office (AMNPO) Intellectual Property (IP) task team was formed to investigate IP matters as they relate to the proposed National Network for Manufacturing Innovation (NNMI or Network) program and develop draft performance-based principles that would govern IP for the NNMI and at the Institutes of Manufacturing Innovation (Institutes) that are created once appropriate legislation has been enacted.[6] The NNMI program intends to establish a network of Institutes intended to anchor a region and the Nation's innovative infrastructure and maximize impact on American manufacturing. This set of principles would allow Institute applicants to propose their plans for how IP rights for a specific Institute would be optimally protected, shared and allocated.

## B. Recommendations

This document addresses key issues associated with IP, especially as related to Institute sustainability and industry impact. The IP provisions of the Bayh-Dole Act are a proven framework for promoting the commercialization and public availability of federally funded research, and are thus the starting point for NNMI IP discussions. Additional considerations are presented here to further promote performance metrics and establish the IP rights of government, Institute and partners, such as:

- IP rights should encourage sustained Institute membership starting from Institute formation, and should discourage members from "fence sitting" or delaying their involvement.
- IP rights should be structured to encourage smaller firms to participate, since small to medium-sized enterprise (SME) involvement in the Institutes is essential.
- IP rights and licensing terms should promote Institute sustainability beyond initial Federal funding.
- Federal Government rights in IP resulting from federally funded research and development awards should be treated in accordance with existing legislation and regulation.

- Publication, data management and export control issues need to be defined.

The draft recommendations herein rely heavily on existing statutes, administrative practices and federal policies regarding funded research. This document makes general recommendations for the program elements related to IP while considering performance metrics. It separates issues into two categories: those that are thought to be important enough to receive attention in the program formulation (Institute Required [7]); and other topics that are thought to be suitable for Institutes to develop their own practices (Institute Discretionary [8]). The draft recommendations are presented in the following table.

| | *Principle* | *Institute* | |
|---|---|---|---|
| | | *Required* | *Discretionary* |
| Institute Rights | 1 *An Institute shall receive a royalty-free, non-exclusive research license to IP generated with Institute or federal funding. Institutes shall have a continued ability to conduct research using such IP for research purposes and can grant commercial license to third parties. Institutes shall determine the terms of distribution of IP (free or royalty bearing license; license limitations; tiered rates or constant; share of royalties to inventors) and negotiate assignment and/or licensing to Institute members. | X | |
| | 2 Institutes should be able to decide for themselves whether any Institute member can provide "in-kind" contributions, including IP, in lieu of membership fees, and should be able to determine the monetary value for such contributions. | | X |
| | 3 The rights and obligations regarding IP sharing and IP ownership of initial members of an Institute as opposed to organizations or companies that join later should be defined by the Institutes | | X |
| | 4 There must be an IP Management Plan submitted to the NNMI program as part of any application to become an Institute or join the NNMI program. | X | |
| | 5 Contents of the IP Management Plan should be determined by the proposers/Institutes. | | X |

| | Principle | Institute Required | Institute Discretionary |
|---|---|---|---|
| | 6 *Upon the dissolution of an Institute, existing IP licenses must be treated according to the particular terms stated in the license agreements and the Institute's IP Management Plan. IP for which title is not owned by the Institute, but which is licensed by the Institute and sublicensed to its member(s) must have the sublicense survive Institute dissolution. Should an Institute cease to exist due to bankruptcy, IP for which title is owned by the Institute must be treated as an asset by a bankruptcy judge. | X | |
| Project Specific | 7 The Institute must be free to pursue non-government funded contract research for industry as a route to sustainability. | X | |
| | 8 The use of government-funded Institute equipment and facilities during research conducted at an Institute solely with industry funds (including full overhead) should not create a government use right or "march in right" for resultant IP. | X | |
| Government Rights and Interests | 9 *The Bayh-Dole Act and regulations will apply to medium and large businesses, and all contractors regardless of type, giving them the same IP ownership election rights (inventor may retain title to generated IP, not funder) as small businesses and non-profits. | X | |
| | 10 Foreign companies may become a member of an Institute and/or may participate in Institute activities when it is in the economic interest of the United States. The national impacts of the proposed membership and/or participation on domestic manufacturing must be evaluated as part of any application to become an Institute or join the NNMI program, or while an Institute is part of the Network. | X | |
| | 11 Impact on domestic manufacturing must be evaluated as part of any application to become an Institute or join the NNMI program. | X | |
| | 12 The language used in the proposal for an Institute addressing impact on domestic manufacturing must be written into the Institute award and be a requirement for eligibility for subsequent competitive grant funding described in the January 2013 NNMI Preliminary Design.[2] | X | |

| | *Principle* | *Institute* | |
|---|---|---|---|
| | | *Required* | *Discretionary* |
| | 13 A Data Management Plan must be submitted that screens companies and data for export control. All proposals must have a plan to comply with export control law. | X | |
| | 14 The Federal government has the right to select any application to become an Institute or join the NNMI network, in whole or in part, as a part of any award or agreement negotiation. | X | |
| | 15 An Institute within the NNMI shall have policies that allow for the results of federally funded research to be made publicly available through publication. However, some data may have significant proprietary value, and it is permitted to require waiting a reasonable period of time before publishing. Each application should establish procedures for publication review prior to publication. | | X |

*Note: It is recognized that some of these principles may conflict with existing technology transfer law and regulations for some Federal agencies, particularly the Bayh Dole Act and its implementing regulations. Appropriate legislative action would be necessary if it were desired to implement those recommendations consistently for all Federal agencies.

## C. Conclusion

The draft IP recommendations that are open for public comment and contained in this document are intended to address the primary IP issues that Institutes created once appropriate legislation has been enacted will need to address. This document presents items of focus for the Federal Government and Institutes, and classifies them as either required or discretionary for the Institutes. With guidance, it is intended that Institutes will develop specific IP rules.

The intent of this draft guidance is to provide great flexibility to the Institutes in designing their IP plans while preserving key IP rights. Chief among these rights are: the government's right to practice IP developed through federally funded research and development awards; an Institute shall receive a royalty-free, non-exclusive research license to IP generated with Institute or federal funding and can grant commercial license to third parties; the right for Institutes to establish a tiered royalty system for members and for non-members (if IP is to be licensed to non-members); the ability of Institutes to enter into contracts to perform research; and the need for evaluation criteria that emphasize American manufacturing impact.

## End Notes

[1] Report available at http://www.whitehouse.gov/sites/default/files/microsites/ostp/pcast-advanced-manufacturingjune2011.pdf.

[2] President Obama to Announce New Efforts to Support Manufacturing Innovation, Encourage Insourcing ; http://www.whitehouse.gov/the-press-office/2012/03/09/president-obama-announce-new-efforts-supportmanufacturing-innovation-en.

[3] "Request for Information on Proposed New Program: National Network for Manufacturing Innovation (NNMI)," 77 FR 26509, May 4, 2012. Available at https://federalregister.gov/a/2012-10809.

[4] Reports are available at http://www.manufacturing.gov/pubs resources.html, under the "Advanced Manufacturing National Program Office (AMNPO)" heading.

[5] Available at http://www.manufacturing.gov/pubs resources.html.

[6] Current proposed legislation includes H.R.2996 entitled "Revitalize American Manufacturing and Innovation Act of 2013," found at http://thomas.loc.gov/cgi-bin/bdquery/z? d113:HR02996:, and the corresponding "Revitalize American Manufacturing and Innovation Act of 2013" introduced on August 1, 2013 by U.S. Sens. Sherrod Brown (D-OH) and Roy Blunt (R-MO).

[7] "Required" corresponds to a principle that is considered critical for Institute sustainability and maximizing the industry impact of resulting technology.

[8] "Discretionary" corresponds to a principle that is recommended but is left to the discretion of each individual Institute to determine if such a principle better serves its members in protecting their IP while maximizing the likelihood of IP commercialization.

In: National Network for Manufacturing … ISBN: 978-1-63117-681-4
Editor: Vivianna R. Bagnoli 

*Chapter 6*

# DRAFT INSTITUTE PERFORMANCE METRICS FOR THE NATIONAL NETWORK FOR MANUFACTURING INNOVATION*

*Edited by Michael F. Molnar*

## FOREWORD

The interagency Advanced Manufacturing National Program Office (AMNPO) is hosted by the National Institute of Standards and Technology (NIST). Creation of the AMNPO flows from the recommendation of the President's Council of Advisors on Science and Technology (PCAST) in its June, 2011, Report to the President on Ensuring American Leadership in Advanced Manufacturing,[1] that the Federal government launch a concerted, whole-of-government Advanced Manufacturing Initiative. To that end, this interagency office is charged with:

- Convening and enabling industry-led, private-public partnerships focused on manufacturing innovation and engaging U.S. universities, and

* This is an edited, reformatted and augmented version of a document released by the Advanced Manufacturing National Program Office, dated November 2013.

- Designing and implementing an integrated whole of government advanced manufacturing initiative to facilitate collaboration and information sharing across federal agencies.

By coordinating federal resources and programs, the AMNPO seeks to enhance technology transfer in U.S. manufacturing industries and help companies overcome technical obstacles to scaling up production of new technologies.

The National Network for Manufacturing Innovation (NNMI) program proposed by President Obama has the goal of advancing American domestic manufacturing.[2] The program will seek to accomplish this by creating a robust national innovation ecosystem anchored by a network of Institutes for Manufacturing Innovation. The NNMI will fill a gap in the innovation infrastructure, allowing new manufacturing processes and technologies to progress more smoothly from basic research to implementation in manufacturing. The NNMI program has a scale and focus that is unique, and it is built upon concepts of a strong public-private partnership.

## ABSTRACT

Beginning in April 2012, a broad public engagement strategy by the Advanced Manufacturing National Program Office (AMNPO) was used to collect extensive input on the National Network for Manufacturing Innovation (NNMI or Network) program design. The collection of information from the public was initiated by a NIST Request for Information (RFI), published in the Federal Register[3], followed by a series of regional workshops sponsored by AMNPO partner agencies and focused on the issues presented in the RFI. Reports summarizing the findings from the RFI and each workshop were published.[4] In January 2013, the *National Network for Manufacturing Innovation: A Preliminary Design* report was published, built upon public input received.[5] This AMNPO document utilizes the information gathered and provides draft performance metrics that can be used to evaluate the performance of Institutes for Manufacturing Innovation (Institutes) within the NNMI program, and that would apply to the NNMI and the Institutes that are created once appropriate legislation has been enacted.[6] This document is one of a series of documents generated to inform and seek feedback from the public on various elements that go into the creation and growth of the NNMI.

More specifically, this document describes draft metrics that can be used to help evaluate the performance and success of Institutes within the NNMI. These draft metrics were developed by an interagency team of experts, building on the initial metrics and lessons learned from the pilot institute, the National Additive Manufacturing Innovation Institute (NAMII). It is recognized that Institutes will vary considerably in their focus areas and perhaps in their structures as well; hence not all of the proposed metrics will be appropriate for all Institutes. However the six categories of draft metrics shown below are considered applicable to all Institutes:

7) Impact (1.1 – 1.4)
8) Industry Value (2.1 – 2.13)
9) Education and Workforce Development (3.1 – 3.6)
10) Portfolio (4.1 – 4.2)
11) Financial (5.1 – 5.5)
12) Network Contribution (6.1 – 6.4)

Within each broad category, several example metrics are given, and a short rationale is included for each. The examples are not intended to be comprehensive, and Institutes should be free to propose other metrics that are applicable to their particular focus areas. Among the example metrics, an effort was made to minimize the reporting burden. Some of the draft metrics are quantitative and some are qualitative. The example metrics also include items that are measurable in short, medium, or long time-frames.

# REPORT

## A. Introduction

The draft performance metrics listed in this report are intended to be used to help measure the successes of the Institutes of Manufacturing Innovation (Institutes) that are a part of the National Network for Manufacturing Innovation (NNMI or Network). The metrics are intended to serve multiple purposes. This includes, as examples, performance measurement of an Institute, self-evaluation that could result in better management, and reports that convey information about Institute and NNMI impacts to a broad audience. Two characteristics can be assigned to each metric. One characteristic is whether the metric is quantitative or qualitative. The second characteristic is the time-frame over which a metric should be assessed. Table

1, at the end of this document, notes both characteristics for all metrics presented.

Institutes will vary considerably in their focus areas and perhaps in their structure, so all metrics may not be appropriate for all Institutes. However, it is anticipated that the six categories of draft metrics will apply to each Institute. The metrics categories are:

1) Impact,
2) Industry Value,
3) Education and Workforce Development,
4) Portfolio,
5) Financial, and
6) Network Contribution.

All Institutes will utilize these six categories of metrics. Some of the draft metrics are designated as "core," indicating that they are common to all Institutes. Other metrics, designated as "example" may or may not be appropriate for a particular Institute, and Institutes could certainly propose alternate metrics of their own. The following Section presents both core and example metrics within all six metric categories.

## B. Institute Metrics

### *1. Impact*

Impact metrics are intended to measure the broad impact of an Institute on U.S. manufacturing. The Institute's impact on manufacturing innovation, employment, and the regional manufacturing ecosystem are particularly of interest.

Core and example Impact metrics include:

### 1.1. Success Stories and Case Studies (Core)

Success stories may be difficult to quantify, but they are an excellent measure of the health of an Institute. The stories can provide a comprehensive look at how the investments in the Institutes and the Network have succeeded, and how the Institutes and the Network have achieved their missions. Success stories will vary between and within Institutes, but they should be easily understood, compelling, and supported by objective data. Success stories may combine information specifically available from other metrics in a way that provides a more complete picture, or they may add information not easily measured elsewhere.

Case studies tell the stories of particular activities undertaken by Institutes to address their missions. Case studies might include items such as:

- Tracking an innovation through the Technology Readiness Levels (TRLs) / Manufacturing Readiness Levels (MRLs) into a commercial product.
    - Identifying manufacturing innovations passing through the Institute
    - Tracking the innovation into the marketplace as a new product or process, including lessons learned; noting what worked and what did not along the path to development (so others can learn from them, too)
- Documenting the impact of the new product or process on competitiveness (for example cost reduction, quality improvement, new market creation, etc.)
    - Envisioning what future uses and impacts might be
    - Documenting changes in technology ecosystems.
    - Companies whose business is within the focus area relocating near the Institute
    - Regional employment rates in the focus area
    - Job creation in the sector beyond the region
    - Company testimonials
    - Federal agency benefits
    - Hiring at regional educational institutions in support of the focus area
    - Technical ecosystem expansion
- Tracking evolving industrial interactions.
    - Identifying the reasons a company became involved with an Institute
    - Tracking the value perceived by partner companies
    - Measuring the number of hires made through the Institute
    - Measuring intellectual property (IP) licensing

## 1.2. Number of Jobs Created and Retained (Core)

As far as is practical, this metric should include only jobs directly attributable to the effects of the Institute. This is a statistic that is more readily available from partner companies, but which may be difficult to quantify beyond the partners. Trends may be available such as

- Regional employment rate
- National employment rate
- Number of graduates from universities, community colleges, and training programs who find employment in the sector

**1.3. Number of Institute Technologies Reaching Commercial Production (Example)**

This metric consists of tabulating and tracking the TRL / MRL of the technology over the life of the Institute. This metric addresses the Institute mission to move innovation from basic research to commercial application.

**1.4. Transitioning Efficiency through the TRL/MRL Levels (Example)**

Some of the technologies that pass through an Institute will reach commercial applications quickly, some more slowly, and some not at all. A high-performance Institute will improve the efficiency at which new technologies progress within and between TRL / MRL levels, reach the marketplace, and compete internationally.

***2. Industry Value***

Industry Value metrics are intended to measure the extent to which the industrial partners perceive that they are receiving value from the existence of the Institute. They measure the industry view of the appropriateness of the selected focus area and of the structure and operation of the Institute.

Core and example Industry Value metrics include:

**2.1. Level and Quality of Co-Investment by Non-Federal Sources (Core)**

This metric enables evaluation of how well the focus area of the Institute matches a real national need. Non-federal partners dedicate resources when they believe that there will be economic benefit. Non-federal sources include cash and in-kind provided by industry partners of all sizes, state and local governments, economic development entities, institutions of higher education, private organizations and individuals.

### 2.2. Trend of Co-Investment by Non-Federal Sources (Core)

In a successful Institute, there must be strong partner co-investment. The co-investment requirement is not trivial, and at the start of an Institute there must be sufficient commitment to warrant the award. Even so, as time goes on, the level of non-federal investment should increase. As the Institute demonstrates success and value, it is expected that new partners will engage, and existing partners will increase their level of engagement.

### 2.3. Ratio of Received to Originally Committed Co-Investment (Example)

In the proposal phase, partners may commit support to an Institute that would be spread over a number of years. The magnitude of these commitments as measured in financial, personnel, and resource services will be as important as the number of the commitments at all stages of an Institute's lifecycle. If an Institute is successful, the ratio of received to originally committed co-investment will grow. If the Institute is not successful, the number of partners and the size of their investments will diminish and the ratio will drop.

### 2.4. Total Number of Partner Companies (Example)

If a broad industrial base recognizes value in an Institute focus area and sees positive impacts from Institute activities, then many companies will want to be partners.

### 2.5. Number of Partner Companies by Size (Small, Medium, and Large) (Example)

Successful Institutes will need partners of all sizes. Historically, many innovations make it to the shop floor or marketplace through the efforts and growth of small and medium-sized SMEs) companies that are very cash limited. Hence the Institutes will need significant and sustained monetary support from large industry members.

### 2.6. Trend in Total Partner Companies (Example)

Partner companies may come and go depending on their financial situations, perceived value from the Institute, personnel changes, etc. In a successful Institute, it is expected that the number of partner companies will increase from initiation but will nominally stabilize over longer time periods.

### 2.7. Growth in Partner Companies by Size (Example)

Growth in a large company could be used, for example, as an indicator of stable and reliable longterm funding. Growth in SMEs could be used as an indicator of dynamic innovation deployment.

### 2.8. Total Number of Retained Partner Companies (Example)

Partners who return with new projects, who continue to provide funding and other resources, who continue to hire from an Institute are an indication of the success of the program. Failure of companies who join at the start to continue their participation could indicate a problem. A reasonable measure might be the yearly ratios of partners from prior years who continue to participate to those who do not.

### 2.9. Number of Retained Partners by Size (Example)

Institutes need to retain participation of large company partners for stability and SMEs for growth and dynamism.

### 2.10. Investment by Partners in Advanced Manufacturing Innovation (Example)

It is expected that Institutes will encourage not only partner investment in the Institute activities, but will lead to increased investment in advanced manufacturing innovation by the partner. Measures that may be used as a proxy of a partner's increased investment in innovation can be improvements in R&D investment, an increase in products developed, and changes in IP developed and/or licensed, as examples.

### 2.11. Number of Companies Making Use of Institute Facilities (Example)

It is expected that Institutes will have unique and advanced facilities compared to their partners and that companies in the ecosystem will want to use the Institute facilities, whether they are partners or not. The uses might include:

- Fee for service
- Participation in training
- Number of startup companies incubating in the Institute.

### 2.12. Number of Spin-Off Companies Created (Example)

Innovations from the Institute could be absorbed by partner companies or may lead to formation of new companies. The latter is particularly true if the innovation is disruptive in some segment. Spin-off companies may provide a means to prove the commercial value of a new innovative technology.

### 2.13. Supply Chain Engagement and Development (example)

A robust supply chain is critical to continued expansion of technology transition. As the number of technologies transitioned increases, it is expected that Institutes will actively seek to establish or expand supply chains. Measurement of supply chain engagement and development is possible. Opportunities exist to partner with the NIST Manufacturing Extension Partnership (MEP) Program (http://www.nist.gov/mep/) to develop metrics in this area. The NIST MEP Program surveys clients to measure jobs created, jobs retained, change in sales, investments leveraged, and cost savings. This knowledge could be expanded to reflect and measure supply chain engagement and development.

## *3. Education and Workforce Development*

The Institutes will have a mission to increase and improve the workforce prepared for advanced manufacturing jobs. This group of metrics is intended to measure success in this mission across a broad spectrum of activities.

Core and example Education and Workforce Development metrics include:

### 3.1. Number of Partner and of Non-Partner Professionals Participating in Research, Education, and Training (Core)

A successful Institute will provide opportunities and programming for education and workforce development. These activities will improve the manufacturing climate in the broader community and draw new partners to the Institute. It is expected that these activities will be interesting not only to partners, but to non-partners as well. Examples of metrics include:

- Number of non-partner attendees at workshops and short courses. The workshops and short courses could be created by Institute staff or they could be related to the focus and simply hosted at the Institute.
- Number of undergraduate students, graduate students, or post-docs drawn from outside the partner institutions to work on the Institute's focus areas.

- Participants from the non-technical community in open-houses, demonstrations, science fairs, Engineer's Day, etc.

### 3.2. Number of University Students Participating in Research, Education, and Training (Core)

It is expected that the Institutes will draw students, particularly from the partner educational institutions, to work in the Institute. Geographical proximity would allow part-time work, student projects, shadowing, and the like. Also, the use of facilities during courses, either in-person or remotely using distance learning technologies, will indicate the educational impacts of the Institute.

### 3.3. Number of Community/Technical College Students Participating in Research, Education, and Training (Example)

It is expected that the Institutes will provide practical exposure to state-of-the-art facilities for community/technical college students, and facilitate pathways for students to learn about educational options that might best suit their long-term interests and capabilities. Institutes and community/technical colleges will promote various technical engagement opportunities including course and work participation. It is expected that institutes will leverage geographic proximity, established regional and professional networks, and forge new relationships.

### 3.4. Number of K-12 Students and Teachers Participating in Research, Education, and Training (Example)

Institutes will have a central role in improving the image of manufacturing. Showcasing the activities of the Institute and partner companies to K-12 students, teachers, parents, and families builds recognition for the interesting, challenging, rewarding careers in manufacturing. It is expected that Institutes will provide easy access for tours, and the Institutes will engage in outreach activities.

### 3.5. Number of Veterans Participating in Research, Education, and Training (Core)

One of the NNMI missions is to tap the talent pool of military veterans. Institutes will provide workforce development programs with specific attention to veteran needs including: certifications, educational opportunities, skills redirection, and others as examples.

### 3.6. Number of Certification and Degree Programs Created in Collaboration with Colleges, Universities, and Professional Organizations (Core)

In addition to providing training and education, Institutes will help develop new and expanded certifications, training programs, degree programs, and other educational opportunities. Institutes might make available examples of course materials that could be developed by Institute members (e.g., lecture materials and homework problems) in the context of undertaking technology development at the Institute. These course materials can be made available to educators to provide real-world content for their curricula.

## *4. Portfolio*

This group of metrics is intended to measure the breadth and depth of projects contained in the Institute portfolio and to track progress toward completion of project objectives.

Core and example Portfolio metrics follow:

### 4.1. Number of Projects in the Portfolio (Core)

A well-functioning Institute will have a portfolio of projects that is broad and deep. Through the life of the Institute, some projects will move out into the commercial world, some will terminate, and new ones will arise. It is expected that the number of projects will grow in the beginning and stabilize at a level that is appropriate for the focus area and size of the Institute.

### 4.2. Number of Project-Level Metrics Achieved (Example)

Each of the projects pursued by an Institute will have measurable outcomes. It is expected that the Institute will make consistent progress in achieving the metrics, or projects should be terminated.

### 4.3. Number and Value of IP Products Produced and Licensed (Core)

This metric includes patents, provisional patents, trade secrets, copyrighted works, and generally, any form of IP. Institutes will be free to set their own IP policies, consistent with the published IP guidance and to use them as a means of encouraging companies to join, especially SMEs. For example, an Institute can consider tiered royalty rates within the Institute (all members pay some, so some funds flow to the inventors and some to help the institute become self-sustaining) and without (non-members pay more for IP).

### *5. Financial*

Institutes need to establish stable revenue streams that will lead them to self-sufficiency after the initial NNMI funding expires. While many long-term funding models are possible, Institutes should demonstrate progress toward self-sufficiency.

Core and example Financial metrics follow:

#### 5.1. Ratio of Membership Dues Income to Institute Expenses (Core)

One source of on-going revenue for an Institute could be dues paid by partner organizations. The ratio of dues income to Institute expenses could be a measure of transition toward sustainability.

#### 5.2. Level of Fees for Services or Publications (Core)

Because the Institute will have unique equipment and capabilities, potential revenue stream could be realized from fees-for-services. The Institute's unique capabilities may also lead to revenue generating publications and fee-based documents.

#### 5.3. Level of Non-Federal Funding (Core)

An Institute may perform work under contracts with both member and non-member organizations, though the terms of the contract may vary depending on the membership status of the organization. In addition, an Institute may be successful in obtaining non-federal funding from various sources through grants or other agreements. Thus the revenue stream an Institute is it able to generate through contracts and non-federal funding will help to measure the Institute's progress toward self-sufficiency.

#### 5.4. Level of Non-NNMI Federal Contracts and Grants (Core)

Federal funding for the NNMI is time limited by design. Institutes are intended to become self-sufficient, meaning not dependent on further NNMI funding. However, Institutes can compete for non-NNMI funding from federal sources. The success of an Institute in securing other federal funding, whether in number of awards or dollars, may be a useful indicator of the value of the Institute to agencies and their missions.

#### 5.5. Level of Intellectual Property (IP) Revenue (Core)

This metric may be stronger for some Institutes, and less so for others. IP revenue can come from: x Direct IP licensing

- The rise in value of a start-up company where the Institute has an equity stake (in exchange for the IP)
- Increased participation in the Institutes by member companies benefiting from the IP produced.

***6. Network Contribution***

Each Institute is a member of the larger NNMI. The interaction of Institutes through the Network will amplify the benefit beyond what any Institute could achieve alone.

Core and example Network Contribution metrics follow:

**6.1. General Interaction with the Larger Network of Institutes (Core)**

A successful Network relies on a multitude of interactions between the individual Institutes and the Network. The interactions include sharing best practices, sharing case studies, building common legal and management frameworks (to the extent possible), and sharing aggregated financial and technical performance information.

**6.2. Number of Referrals of Projects or Partners to Other Institutes in the Network (Example)**

Institutes will have an understanding of unique skills, capabilities, and services other Institutes possess within the Network, and they therefore can serve an important role in referring potential partners to the most appropriate Institute for their interests. The number of referrals received and that lead to substantive joint activities within the Institute would be a measure of Institute and Network performance.

**6.3. Number of Projects or Partners Received from Other Institutes in the Network (Example)**

By strongly communicating their focus areas to other Institutes, an Institute can better attract referrals from across the Network.

**6.4. Institute Participation in Network Governance (Core)**

Institute participation in Network activities on a regular basis will strengthen the Network, may strengthen the performance of the Institute, and would be an indicator of the Institute's leadership. Consequently, contributions made by an Institute to the governance of the Network should be assessed using qualitative and quantitative methods.

## Table 1. Classification of Metrics

| Metric Classification | Short Time Frame | Medium Time Frame | Long Time Frame |
|---|---|---|---|
| Qualitative | 1.1 2.13 4.2 6.1<br>6.4 | 1.1 2.13 4.2 6.1 | 1.1 4.2 6.1 |
| Quantitative | 4.1<br>2.1 3.1 5.1<br>4.3<br>2.3 3.2<br>5.3<br>2.4 3.3<br>2.5 3.4<br>2.10 3.5<br>3.6 | 1.2 2.2 3.1 4.1 5.1 6.2<br>1.3 2.3 3.2 4.3 5.2 6.3<br>2.4 3.3 5.3 6.4<br>2.5 3.4 5.4<br>2.6 3.5 5.5<br>2.7 3.6<br>2.8<br>2.9<br>2.10<br>2.11<br>2.12<br>2.13 | 1.2 2.2 3.1 4.1 5.1 6.2<br>1.3 2.3 3.2 4.3 5.2 6.3<br>1.4 2.4 3.3 5.3 6.4<br>2.5 3.4 5.4<br>2.6 5.5<br>2.7<br>2.8<br>2.9<br>2.10<br>2.11<br>2.12<br>2.13 |

## End Notes

[1] Report available at http://www.whitehouse.gov/sites/default/files/microsites/ostp/pcast-advanced-manufacturingjune2011.pdf.

[2] President Obama to Announce New Efforts to Support Manufacturing Innovation, Encourage Insourcing, http://www.whitehouse.gov/the-press-office/2012/03/09/president-

[3] "Request for Information on Proposed New Program: National Network for Manufacturing Innovation (NNMI)," 77 FR 26509, May 4, 2012. Available at https://federalregister.gov/a/2012-10809.

[4] Reports are available at http://www.manufacturing and are listed under the "Advanced Manufacturing National Program Office (AMNPO)" heading.

[5] Available at http://www.manufacturing resources.html.

[6] Current proposed legislation includes: H.R.2996 entitled *"Revitalize American Manufacturing and Innovation Act of 2013,"* found at http://thomas.loc.gov/cgi-bin/bdquery/z?d113:HR02996:, and the corresponding *"Revitalize American Manufacturing and Innovation Act of 2013"* introduced on August 1, 2013 by U.S. Sens. Sherrod Brown (D-OH) and Roy Blunt (R-MO).

# INDEX

## D

## E



## F



## G

## H



## I



## J

## K



## L



## M



## N

## O



## P



## Q



## R

## S

## T



## U



## V



## W

## Y